Practical Applications in Reliability Engineering

Edited by Muhammad Zubair

Published in London, United Kingdom

IntechOpen

Supporting open minds since 2005

Practical Applications in Reliability Engineering
http://dx.doi.org/10.5772/intechopen.91570
Edited by Muhammad Zubair

Contributors
Ibrahim Yusuf, Ismail Muhammad Musa, Muhammad Zubair, Priyonta Rahman, Ibrahima Dit Bouran Sidibe, Djelloul Imene, Abdou Fane, Amadou Ouane, Viera Stopjakova, Michal Sovcik, Lukas Nagy, Daniel Arbet, Eslam Ahmed

Notice
Statements and opinions expressed in the chapters are these of the individual contributors and not necessarily those of the editors or publisher. No responsibility is accepted for the accuracy of information contained in the published chapters. The publisher assumes no responsibility for any damage or injury to persons or property arising out of the use of any materials, instructions, methods or ideas contained in the book.

First published in London, United Kingdom, 2021 by IntechOpen
IntechOpen is the global imprint of INTECHOPEN LIMITED, registered in England and Wales, registration number: 11086078, 5 Princes Gate Court, London, SW7 2QJ, United Kingdom
Printed in Croatia

British Library Cataloguing-in-Publication Data
A catalogue record for this book is available from the British Library

Additional hard and PDF copies can be obtained from orders@intechopen.com

Practical Applications in Reliability Engineering
Edited by Muhammad Zubair
p. cm.
Print ISBN 978-1-83968-399-2
Online ISBN 978-1-83968-400-5
eBook (PDF) ISBN 978-1-83968-401-2

We are IntechOpen,
the world's leading publisher of
Open Access books
Built by scientists, for scientists

5,300+
Open access books available

130,000+
International authors and editors

155M+
Downloads

Our authors are among the

156
Countries delivered to

Top 1%
most cited scientists

12.2%
Contributors from top 500 universities

Interested in publishing with us?
Contact book.department@intechopen.com

Numbers displayed above are based on latest data collected.
For more information visit www.intechopen.com

Meet the editor

Dr. Muhammad Zubair is an associate professor in the Department of Mechanical and Nuclear Engineering, University of Sharjah, United Arab Emirates. Prior to this role, Dr. Zubair worked as an assistant professor and graduate program coordinator at the University of Engineering and Technology Taxila, Pakistan. Dr. Zubair's interests include nuclear reactor safety, accident analysis, reliability and risk analysis, digital instrumentation and control, and radiation detection and measurements. He has a strong research background supported by publications in international journals, conferences, and book chapters. He is engaged in different research projects including one coordinated by the International Atomic Energy Agency (IAEA). He also serves as editor, associate editor, and technical committee member for international journals and conferences.

Contents

Preface

This book presents the state of the art in reliability and risks analysis engineering from a product lifecycle standpoint. The book provides comprehensive insights into optimization of systems, digital and analog systems, uncertainty quantification, and maintenance as well as risk analysis. The book is intended for senior undergraduate and postgraduate students in related engineering programs such as mechanical engineering, manufacturing engineering, industrial engineering, and engineering management programs, as well as for researchers and engineers in the reliability and risk fields.

The book is structured in a way that the optimization methods for components and systems are described first. This is because components are the most basic building blocks of engineering systems. The techniques discussed in this chapter provide a comprehensive overview of some reliability features of series-parallel systems under minor and complete failure. Failure and repair time of all units were assumed to be exponentially distributed. Explanatory expressions for system characteristics such as system availability, mean time to failure (MTTF), and profit function and cost benefits for all configurations have been obtained and validated by performing numerical experiments. The next chapter describes reliability analysis in the nuclear industry. It investigates the Containment Spray System (CSS) by using the fault tree analysis (FTA) technique to obtain results of the top event probabilities, minimal cut sets (MCS), risk decrease factor (RDF), fisk increase factor (RIF), and sensitivity analysis.

The chapter on maintenance and uncertainty addresses a maintenance optimization problem for remanufactured equipment that will be reintroduced into the market as secondhand equipment. It discusses and derives an optimal maintenance policy for such equipment to overcome the uncertainty of reparation action. Moreover, the chapter presents experiments and evaluates different life cycles of technologies according to their obsolescence processes (accidental or progressive vanishing) on the optimal operating condition. The last chapter on reliability for digital and analog systems deals with a digital method of calibration for analog-integrated circuits as a means of extending its lifetime and reliability, which consequently affects the reliability of the analog electronic system. The chapter reveals the implementation of an ultra-low voltage on-chip system of the digitally calibrated variable-gain amplifier (VGA), fabricated in CMOS 130 nm technology.

The editor and the contributors (authors) would like to thank Ms. Dolores Kuzelj and the staff at IntechOpen.

Muhammad Zubair
Department of Mechanical and Nuclear Engineering,
University of Sharjah,
Sharjah, United Arab Emirates

Section 1

Introduction

Chapter 1

Introductory Chapter: An Overview of Reliability and Risk Analysis

Muhammad Zubair and Eslam Ahmed

1. Introduction

Reliability improvement can be acquired through such measures as testing, periodic examinations, support, and quality assurance for exercises influencing the quality of a nuclear power plant. Reliability engineering can add to these actions through proceeded with assessment of the viability with which assets are applied to accomplish expressed destinations and exhibit of how they can prompt the advancement of operation and maintenance. In this way, it has been shown that by utilizing disappointment and fix information, one can infer, by use of reliability examination methods, an ideal occasional testing or assessment recurrence, maintenance system, and operation practices. For more extensive use of the methods of reliability engineering in functional plant operation and maintenance, the primary prevention is the truth that these strategies are very novel to the reasonable specialist. Likewise, practical engineers are to some degree less slanted to see the value in the immediate benefits of this methodology in light of the fact that the reliability examiners are sometimes not ready to exhibit that the real presence of the investigation helps design, maintenance, and operational engineers to settle on reasonable choices [1].

2. Quality assurance and quality control

The more modern advancements become, the more significant are quality and reliability perspectives for ensuring the properties and operational attributes of the innovation. This is especially valid for enterprises such as nuclear energy, which are conceivably hazardous for individuals and the environment because of the utilization of radioactive materials and highly concentrated energy density. At the point when applied to nuclear fuel designing, quality assurance and quality control (QA/QC) and reliability necessities are totally interconnected. Notwithstanding, the terms are ideally utilized independently by fuel makers (weight on 'quality') and fuel operators (weight on 'reliability'). The QA/QC techniques and guidelines are a piece of the generally integrated management system (IMS) for an association.

Nuclear power has a place with a profoundly cutthroat power industry that aims for better business nuclear power plant execution inside characterized safety edges. Nuclear power improvement mirrors the advancing trade-off between techno-economic motivations and safety prerequisites. Henceforth, both specialized and safety viewpoints are to be viewed along with administrative methodologies focused on practical, commonsense execution of these substitute inspirations.

A nuclear reactor is by and large described by testing operational conditions, with the most extreme conditions in the reactor core, where high temperatures, corrosive media, and mechanical stresses are joined with concentrated radiation load on fuel elements, fuel assemblies, and reactor internals. These operational angles can prompt the corruption of material properties and eventually to failures of fuel and other reactor internals. The expense of such failures is high, and their outcomes can be amazingly extreme. Hence, careful consideration is given to the appropriate determination, improvement, design, assembling, testing, and operation of fuels and in-core components of nuclear reactors.

While different specialized, safety, and managerial aspects of fuel designing and execution are inspected in various publications, there is a lack of comprehensive direction over the scope of interconnected issues of fuel quality and reliability [2].

3. Risk management

In the current worldwide energy environment, nuclear power plant (NPP) supervisors need to think about numerous hazard components notwithstanding the nuclear safety-related risk. To remain cutthroat in current energy markets, NPP administrators should coordinate management of creation, safety-related, and economic risks compellingly. This risk management (RM) approach produces benefits that incorporate the following: Clearer rules for decision making. Utilizing ventures previously made in probabilistic safety analysis (PSA) programs by applying these examinations to different zones and settings. Cost consciousness and advancement in accomplishing nuclear safety and creation objectives. Correspondence improvement more successful inner correspondence among all levels of the NPP working association, and more clear correspondence between the association and its partners. Focus on safety, guaranteeing an incorporated spotlight on safety, production, and financial aspects during seasons of progress in the energy environment.

Throughout the most recent decade, in the various Member States, there has been a move from nationalized responsibility for utilities inside economies outfitted towards complete and stable business to privatized, serious business sectors with strain to diminish costs, staff numbers, and the designing responsibility. The emphasis presently is on gathering the objectives set by investors instead of governments. Some Member States have not seen such stamped changes, be that as it may, these shifts are characteristic of the bearing of the world's energy markets.

To get by in this new de-regulated and cutthroat environment, NPPs need to protect and keep up safety and focus on market costs, market interest, and execution. Plainly, deregulation builds hazards yet additionally produces openings for more substantial benefits. It is in this setting that NPP operators need to think about all parts of hazard and concoct an ideal arrangement that doesn't bargain safety and execution.

One of the significant advantages of a coordinated risk management approach is that safety, operational, and financial execution (and risks) are frequently connected. NPPs with outstanding safety records will, in general, show solid economic execution, and the other way around [3].

The objective of an integrated risk management approach is to fuse into the association's management framework a structure for a methodical investigation that shows identification and the executives of hazard in a portfolio setting. This incorporated (or portfolio) way to deal with hazard investigation can assist the association with deciding the right blend of preventive measures, transfer of risk to other gatherings, and maintenance of hazard by the association. The advantages will accumulate to the partners, including business or government proprietors and society [4].

Critical infrastructure systems (CISs), for example, nuclear power plants (NPPs) and help organizations, are the foundation of the cultured countries; they give the fundamental energy assets to networks. Notwithstanding, these CISs are frequently inclined to more than a solitary risk. Given the characteristic relationship of natural hazards or unintentionally, CISs can all the while being exposed to multi-hazards, which are simultaneous and progressive events of more than one risk. Multi hazards can additionally build the catastrophe hazard of CISs; be that as it may, contrasted and single hazard risk assessment, multi-hazard risk evaluation is generally new in many exploration spaces [5]. As of late, in any case, a notable multi-hazard occasion, the core damage accident of the Fukushima-Daiichi NPP in March 2011, drove the importance of doing essential multi-hazard risk evaluation. Under these conditions, the endeavors to comprehend and evaluate the multi-hazard chances have expanded in different exploration fields, including geophysics, sociology, underlying designing, reliability engineering, and nuclear safety engineering.

Especially in the scope of nuclear safety designing, the multi-hazard risk should be counted during NPP safety assessment. Albeit the multi-hazard force and its impact can generally be inconsequential under a specific return period chosen by the current design standard, the absolute multi-hazard risk, convolution of yearly event probability, and the result can be non-insignificant in the hazard assessment phase. In contrast to the planning stage, the risk assessment technique incorporates the disproportional results, which are expected under a multi-hazard force that is past the design criteria. The International Atomic Energy Agency (IAEA) distributed a progression of reports (2011, 2017, and 2018) on probabilistic safety assessment (PSA) for NPP multi-hazards. Site-explicit outer risks, external hazard combinations, just as critical structures, systems, and components (SSCs) exposed to multi-hazards, were examined in these reports.

4. Regulatory authorities

Notwithstanding IAEA, the U.S. Department of Energy (USDOE) likewise featured the significance of multi-hazard evaluation for NPP facilities [6, 7]. The progressing venture of the Korea Atomic Energy Research Institute (KAERI), called the "Development of multi natural hazard risk assessment," likewise upholds various multi-hazard research themes, including different multi-hazard combinations (e.g., earthquake mainshock-aftershock, typhoon-earthquake, earthquake-landslide, and earthquake-tsunami) to work with the multi-hazard risk measurement for NPPs [8–10]. In any case, despite the arising need for multi-hazard investigation for NPP systems, the overall strategy is not broadly examined. Contrasted and single hazard risk evaluation, multi-hazard hazard assessment is generally new in the field, and the essential phrasings actually should be characterized.

Accordingly, we expected to survey cutting-edge research in multi-hazard investigation past nuclear safety engineering (e.g., geophysics, structural engineering, reliability engineering) and examine the advancement and difficulties in its application to NPP systems. The fundamental conversation subjects of this investigation are fourfold: order of multi-hazard interaction; the best multi-hazard examination for each multi-hazard combination; the advancement, potential, and difficulties in the use of the momentum multi-hazard examination strategies to NPP constructions and systems; and the flow research holes in the multi-hazard riskevaluation system. Mainly, writing on the state of the craftsmanship, the multi-hazard investigation is discussed as far as risk, delicacy, and hazard examination level. For quantitative evaluation of the multi-hazard risk, both hazard and delicacy models ought to be created, where the delicacy model is the restrictive prospect of a predetermined damage state (e.g., moderate, extensive, or total failure) for a given risk force (e.g., peak ground acceleration, wind speed) [11]. The current advancement phase of the hazard and fragility examination straightforwardly influences the last multi-hazard risk, but it does not really ensure the accessibility of the multi-hazard risk assessment.

In utilizing hazard-educated methodologies for guaranteeing safety regarding working nuclear power plant (NPPs), hazard significance measures got from proba-bilistic risk assessments (PRAs) of the plants are essential components of thought much of the time. Getting these actions in suitable structures is helpful for leaders and can work with the utilization of hazard data.

In this monograph, the emphasis is on hazard significance as evaluated by the PRA models of NPPs created as per current guidelines and devices. The idea of hazard significance measure in PRA is, in numerous applications, identified with a solitary "basic event" and this is the thing that is generally determined by the PRA devices (albeit some of them, like RiskSpectrum, incorporate certain high-level choices, as examined later). Then again, what is of interest in useful applications is the hazard significance of specific segments like pump or valve, which is in cur-rent PRA models ordinarily addressed by different essential occasions where every primary event is identified with explicit failure mod or reason for inaccessibility [12] A similar failure mode may, because of various limit conditions, be introduced by various basic events in various accident arrangements. To convolute the things further, failure modes might be shared by different segments; for example, agent basic event might be an individual from some common cause failure (CCF) group. To plan the significance of specific basic events into the significance of part, some PRA applications, talked about underneath, set up a set of rules to be utilized for the reason. This cycle is relatively convoluted, is dependent upon interpretation, and now and again requires extra assessments. Accessibility of measures that can be straightforwardly associated with a segment of a safety system, "component level" significance measures, can work on the utilization of these actions in numerous applications [13].

Author details

Muhammad Zubair* and Eslam Ahmed
Department of Mechanical and Nuclear Engineering, University of Sharjah, Sharjah, UAE

*Address all correspondence to: mzubair@sharjah.ac.ae

IntechOpen

References

[1] IAEA. Reliability of Nuclear Power Plants. International symposium on reliability of nuclear power plants. (1975).

[2] IAEA. Quality and reliability aspects in nuclear power reactor fuel engineering. IAEA nuclear energy series. (2015). No. NF-G-2.1.

[3] Zubair, M. Ishag, A. Sensitivity analysis of APR-1400's Reactor Protection System by using RiskSpectrum PSA. Nuclear Engineering and Design, Volume 339, pp 225-234, 2018.

[4] IAEA. Risk management: A tool for improving nuclear power plant performance. IAEA nuclear energy series. (2001). IAEA-TECDOC-1209.

[5] Komendantova, N., Mrzyglocki, R., Mignan, A., Khazai, B., Wenzel, F., Patt, A., & Fleming, K. Multi-hazard and multi-risk decision-support tools as a part of participatory risk governance: Feedback from civil protection stakeholders. International Journal of Disaster Risk Reduction. (2014), 8, 50-67.

[6] Coleman, J. L., Bolisetti, C., Veeraraghavan, S., Parisi, C., Prescott, S. R., & Gupta, A. Multi-hazard advanced seismic probabilistic risk assessment tools and applications. (2016a).

[7] Coleman, J. L., Smith, C. L., Burns, D., & Kammerer, A. Development Plan for the External Hazards Experimental Group Report. (2016b).

[8] Hur, J., Shafieezadeh, A. Multi-hazard probabilistic risk analysis of off-site overhead transmission systems. (2019).

[9] Kim, JH., Kim, MK., Choi, IK. Preliminary Study on the Quantification of Component Level Failure Frequency by Multi-Hazard. (2019).

[10] Mun, CU. Bayesian Network for Structures Subjected to Sequence of Main and Aftershocks. Seoul National University. (2019).

[11] Gidaris, I., Padgett, J. E., Barbosa, A. R., Chen, S., Cox, D., Webb, B., & Cerato, A. Multiple-hazard fragility and restoration models of highway bridges for regional risk and resilience assessment in the United States: state-of-the-art review. Journal of Structural Engineering. (2017). 143(3).

[12] Zubair M, Zhang Z. Reliability Data Update Method (RDUM) based on living PSA for emergency diesel generator of Daya Bay nuclear power plant. Safety Science, Volume 59, PP 72-77, 2013.

[13] Vrbanic, I., Samanta, P. Risk Importance Measures in the Design and Operation of Nuclear Power Plants. Brookhaven National Laboratory. (2017).

Reliability for Optimal Systems

The Optimal System for Complex Series-Parallel Systems with Cold Standby Units: A Comparative Analysis Approach

Ibrahim Yusuf and Ismail Muhammad Musa

Abstract

The purpose of this research is to propose three reliability models (configurations) with standby units and to study the optimum configuration between configurations analytically and numerically. The chapter considered the need for 60 MW generators in three different configurations. Configuration 1 has four 15 MW primary units, two 15 MW cold standby units and one 30 MW cold standby unit; Configuration 2 has three 20 MW primary units, three 20 cold standby units; Configuration 3 has two 30 MW primary units and three 30 MW cold standby units. Some reliability features of series–parallel systems under minor and complete failure were studied and contrasted by the current. Failure and repair time of all units is assumed to be exponentially distributed. Explanatory expressions for system characteristics such as system availability, mean time to failure (MTTF), profit function and cost benefits for all configurations have been obtained and validated by performing numerical experiments. Analysis of the effect of different system parameters on the function of profit and availability has been carried out. Analytical comparisons presented in terms of availability, mean time to failure, profit function and cost benefits have shown that configuration 3 is the optimal configuration. This is supported by numerical examples in contrast to some studies where the optimal configuration of the system is not uniform as it depends on some system parameters. Graphs and sensitivity analysis presented reveal the analytical results and accomplish that Configuration 3 is the optimal in terms of design, reliability physiognomies such as availability of the system, mean time to failure, profit and cost benefit. The study is beneficial to engineers, system designers, reliability personnel, maintenance managers, etc.

Keywords: optimality, availability, standby, partial, complete failure, MTTF

1. Introduction

Systems or configurations are designed with intention of meeting the optimal designed that has the reliability requirement at satisfaction of the buyers or customers usually studied with intention to the increase their reliability characteristics in terms such as mean time to failure (MTTF), busy period of repairman, availability, generated revenue as well as profit. Reliability models are vital in measuring the overall performance of system in ensuring quality of products. Achieving a high level of reliability through redundancy is often an essential requisite.

Literature on the reliability of comparative analysis of systems with standby units is numerous, and here we study previous papers on the issues of systems with standby units. Due to their significance in education, communication, military, industry and economics, many researchers have done excellent work in the field of reliability and performance analysis of serial systems by studying and constructing mathematical models to test their performance under different operating conditions. For instance; Singh et al. [1, 2] used copula to study the performance analysis of the complex system in the series configuration under different failure and repair discipline. Lado and Singh [3] recently discussed the cost assessment of complex repairable systems consisting two subsystems in series configuration using Gumbel Hougaard family copula. Yusuf [4] presented the availability modeling and evaluation of repairable system subject to minor deterioration under imperfect repairs. Singh and Ayagi [5] provided a frame work to analyze the performance of a complex system under preemptive resume repair policy using copula. Niwas and Garg [6] discussed the availability, reliability and profit of an industrial system based on cost free warranty policy. Monika et al. [7] provided a complex system having two subsystems in series configuration under k-out-of-n: G, policy. The k-out-of-n works if and only if at least k of the n components works. Gahlot et al. [8] analyzed the performance of repairable system in series configuration under different types of failure and repair policies using copula linguistics. Singh.,V.V and Singh, N. P [9] analyzed the performance of three-unit redundant system with switch and human failure. Saini and Kumar [10] discussed the performance evaluation of evaporation system in sugar industry using RAMD analysis. Malik and Tewari [11] presented performance modeling and maintenance priorities decision for water flow system of a coal based thermal power plant. Lado et al. [12] discussed the performance and cost assessment of repairable complex system with two subsystems connected in series configuration.

Researchers in the past have presented excellent works on reliability analysis of complex repairable systems and proclaimed better performance of the repairable system by their operations. Chen et al. [13] dealt with reliability analysis of a cold standby system with imperfect repair and under poisson shocks. Corvaro et al. [14] presented RAM analysis on reciprocating compressors. Garg [15] analyzed the reliability of industrial system using fuzzy kolmogrov's differential equations. Garg [16] presented an approach for analyzing the reliability of series–parallel system using credibility theory and different types of intuitionistic fuzzy numbers. Garg and Sharma [17] discussed two phase approach for reliability and maintainability analysis of an industrial system. Garg [18] presented RAM analysis of industrial systems using PSO and fuzzy methodology. Kakkar et al. [19] analyzed the reliability of two-unit parallel repairable industrial system. Kakkar [19] discussed the reliability of two dissimilar parallel unit repairable system with failure during preventive maintenance. Niwas and Kadyan [20] dealt with reliability modeling of a maintained system with warranty and degradation. Negi and Singh [21] analyzed the reliability of non-repairable complex system with weighted subsystems connected in series. Patil et al. [22] presented the reliability analysis of CNC turning center based on the assessment of trends in maintenance data. Tsarouhas [23] dealt with RAM analysis for wine packaging production line. Wang et al. [24] analyzed the reliability of two-dissimilar-unit warm standby repairable system with priority in use. Wu [25] analyzed the reliability of a cold standby system attacked by shocks. Wu and Wu [26] analyzed the reliability of two-unit cold standby repairable systems under Poisson shocks. Garg [27] analyzed the reliability of industrial system using fuzzy kolmogrov's differential equations. Kakkar et al. [28] analyzed the reliability of two dissimilar parallel unit repairable system with failure during preventive maintenance. Kumar and Malik [29, 30] dealt with reliability measures of a computer system with priority to PM over the H/W repair activities subject to MOT and MRT. Kumar and Lather [31] analyzed the reliability of a robotic system using hybridized technique.

Kumar et al. [32] dealt with availability and cost analysis of an engineering system involving subsystems in series configuration. Suleiman et al. [33] dealt with comparative analysis between four dissimilar solar panel configurations.

Still, a further study om serial system of the new type of models with a justified and satisfactory assessment is required. For this reason, this chapter has three goals. The first goal is, to develop explicit expressions describing mean time to failure. The second is to compare the four configurations in terms of their mean time to failure. The third is to perform a parametric investigation of various system parameters with the mean time to failure, as well as to capture their effect on the mean time to failure. Analytical and numerical computations are presented to compare their mean time to failure (MTTF). Cost/benefit measure have been obtained for all configurations, where the benefit is mean time to failure.

The rest of the paper is organized as follows. Section 2 presents the notation used. Section 3 gives a description of the system. Section 4 deals with derivation of the models. Analytical comparison between configurations are presented in Section 4. The results of our numerical simulations are presented and discussed in Section 5. The paper is concluded in Section 6.

2. Notations

α_0/β_0: Unit failure/Repair rate.

$p_i(t)$: Probability that Configuration 1/Configuration 2/Configuration 3 is in state i at time t.

$P(t)$: Probability row vector.

$Q_n/A_{Tn}/MTTF_n, n = 1, 2, 3$: Transition matrix/steady state Availability/Mean time to failure for the Configuration 1/Configuration 2/Configuration 3.

$C_1/C_2/C_3$: cost for Configuration 1/Configuration 2/Configuration 3.

$k_0/k_1/k_2$: Revenue generated/cost due to repair of partial failure/cost due to repair of complete failure.

3. Description of the systems

The present paper considered the requirement of 60 MW generators in following configurations: Configuration 1 has four 15 MW primary units, two 15 MW cold standby units and one 30 MW cold standby unit; Configuration 2 has three 20 MW primary units, three 20 cold standby units; Configuration 3 has two 30 MW primary units and three 30 MW cold standby units. It is assumed that units fail independent of the other (**Table 1**). It is also assumed that switching from standby to operation is automatic. Primary unit fails with exponential failure time distribution with parameter α_0 and immediately the cold standby is switch to operation. Also, unit fails independent of the other with exponential failure time distribution with parameter α_0. Both units have exponential repair time distribution with parameter β_0. The systems (Configurations) are depicted in **Figures 1–3** below.

Configuration	Number of Primary units	Number of standby units	Cost of Configuration
1	Four primary 15 MW	Two cold standby 15 MW units	$C_1 = 48,000,000$
2	Three primary 20 MW	Three cold standby 20 MW	$C_2 = 42,000,000$
3	Two primary 30 MW	Three cold standby 30 MW	$C_3 = 39,000,000$

Table 1.
Size of configurations and their corresponding cost.

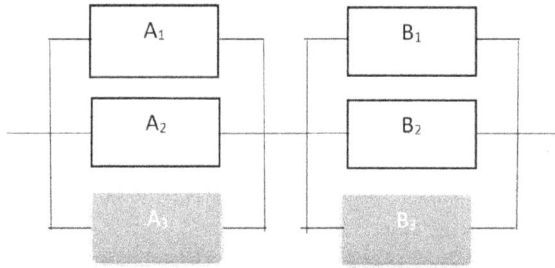

Figure 1.
Reliability block diagram of configuration 1.

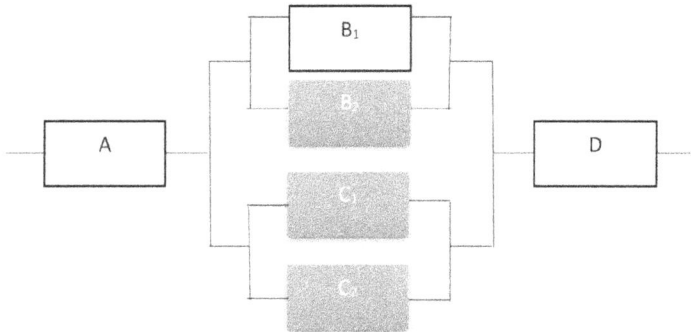

Figure 2.
Reliability block diagram of configuration 2.

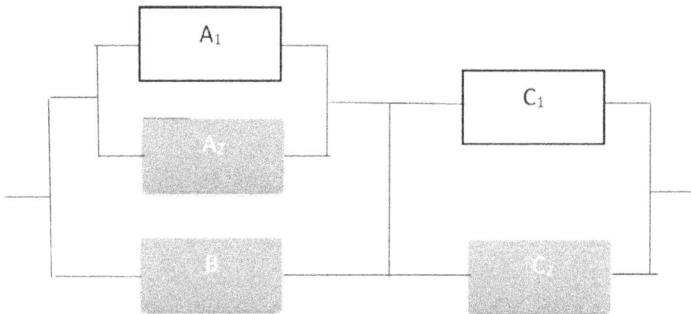

Figure 3.
Reliability block diagram of configuration 3.

4. Reliability models formulation

4.1 Models formulation for configuration 1

The corresponding set of differential-difference equations for Configuration 1 as follows:

$$\frac{d}{dt}p_0(t) = -4\alpha_0 p_0(t) + \beta_0 p_1(t) + \beta_0 p_2(t)$$

$$\frac{d}{dt}p_1(t) = -(4\alpha_0 + \beta_0)p_1(t) + 2\alpha_0 p_0(t) + \beta_0 p_3(t) + \beta_0 p_5(t)$$

$$\frac{d}{dt}p_2(t) = -(4\alpha_0 + \beta_0)p_2(t) + 2\alpha_0 p_0(t) + \beta_0 p_3(t) + \beta_0 p_4(t)$$

$$\frac{d}{dt}p_3(t) = -(4\alpha_0 + \beta_0)p_3(t) + 2\alpha_0 p_1(t) + 2\alpha_0 p_2(t) + \beta_0 p_6(t) + \beta_0 p_7(t)$$

$$\frac{d}{dt}p_4(t) = -\beta_0 p_4(t) + 2\alpha_0 p_2(t)$$

$$\frac{d}{dt}p_5(t) = -\beta_0 p_5(t) + 2\alpha_0 p_1(t)$$

$$\frac{d}{dt}p_6(t) = -\beta_0 p_6(t) + 2\alpha_0 p_3(t)$$

$$\frac{d}{dt}p_7(t) = -\beta_0 p_7(t) + 2\alpha_0 p_3(t)$$

$$\tag{1}$$

With initial conditions

$$p_k(0) = \begin{cases} 1, & k = 0 \\ 0, & k = 1, 2, 3, \dots, 7 \end{cases} \tag{2}$$

Eq. (1) can be expressed in the form as:

$$\frac{d}{dt}p(t) = Q_1 p(t) \tag{3}$$

With

$$Q_1 = \begin{pmatrix}
-4\alpha_0 & \beta_0 & \beta_0 & 0 & 0 & 0 & 0 & 0 \\
2\alpha_0 & -(4\alpha_0 + \beta_0) & 0 & \beta_0 & 0 & \beta_0 & 0 & 0 \\
2\alpha_0 & 0 & -(4\alpha_0 + \beta_0) & \beta_0 & \beta_0 & 0 & 0 & 0 \\
0 & 2\alpha_0 & 2\alpha_0 & -(4\alpha_0 + 2\beta_0) & 0 & 0 & \beta_0 & \beta_0 \\
0 & 0 & 2\alpha_0 & 0 & -\beta_0 & 0 & 0 & 0 \\
0 & 2\alpha_0 & 0 & 0 & 0 & -\beta_0 & 0 & 0 \\
0 & 0 & 0 & 2\alpha_0 & 0 & 0 & -\beta_0 & 0 \\
0 & 0 & 0 & 2\alpha_0 & 0 & 0 & 0 & -\beta_0
\end{pmatrix}$$

Expression of availability, probability of partial and complete failure for configuration 1 are given by

$$A_{T1}(\infty) = p_0(\infty) + p_1(\infty) + p_2(\infty) + p_3(\infty) \tag{4}$$

$$B_{P1}(\infty) = p_1(\infty) + p_2(\infty) + p_3(\infty) \tag{5}$$

$$B_{P2}(\infty) = p_4(\infty) + p_5(\infty) + p_6(\infty) + p_7(\infty) \tag{6}$$

To obtained (4), the procedure is to compute the states probabilities $p_k(\infty), k = 0, 1, 2, \dots, 7$ by setting (3) to zero to give

$$Q_1 P(t)^T = 0 \tag{7}$$

and using the following normalizing condition

$$p_0(\infty) + p_1(\infty) + p_2(\infty) + p_3(\infty) + p_4(\infty) + p_5(\infty) + p_6(\infty) + p_7(\infty) = 1 \quad (8)$$

to give

$$
\begin{pmatrix}
-4\alpha_0 & \beta_0 & \beta_0 & 0 & 0 & 0 & 0 & 0 \\
2\alpha_0 & -(4\alpha_0 + \beta_0) & 0 & \beta_0 & 0 & \beta_0 & 0 & 0 \\
2\alpha_0 & 0 & -(4\alpha_0 + \beta_0) & \beta_0 & \beta_0 & 0 & 0 & 0 \\
0 & 2\alpha_0 & 2\alpha_0 & -(4\alpha_0 + 2\beta_0) & 0 & 0 & \beta_0 & \beta_0 \\
0 & 0 & 2\alpha_0 & 0 & -\beta_0 & 0 & 0 & 0 \\
0 & 2\alpha_0 & 0 & 0 & 0 & -\beta_0 & 0 & 0 \\
0 & 0 & 0 & 2\alpha_0 & 0 & 0 & -\beta_0 & 0 \\
1 & 1 & 1 & 1 & 1 & 1 & 1 & 1
\end{pmatrix}
\begin{pmatrix}
p_0(\infty) \\ p_1(\infty) \\ p_2(\infty) \\ p_3(\infty) \\ p_4(\infty) \\ p_5(\infty) \\ p_6(\infty) \\ p_7(\infty)
\end{pmatrix}
=
\begin{pmatrix}
0 \\ 0 \\ 0 \\ 0 \\ 0 \\ 0 \\ 0 \\ 1
\end{pmatrix}
$$

$$(9)$$

By solving the system of equations in (9) using MATLAB package for the solution of $p_k(\infty)$ give in **Table 2** below.

(4), (5) and (6) are now expressed as:

$$A_{T1}(\infty) = \frac{\beta_0^3 + 4\alpha_0\beta_0^2 + 4\alpha_0^2\beta_0}{16\alpha_0^3 + 12\alpha_0^2\beta_0 + 4\alpha_0\beta_0^2 + \beta_0^3} \quad (10)$$

$$B_{P1}(\infty) = \frac{4\alpha_0\beta_0(\alpha_0 + \beta_0)}{16\alpha_0^3 + 12\alpha_0^2\beta_0 + 4\alpha_0\beta_0^2 + \beta_0^3} \quad (11)$$

$$B_{P2}(\infty) = \frac{8\alpha_0^2(2\alpha_0 + \beta_0)}{16\alpha_0^3 + 12\alpha_0^2\beta_0 + 4\alpha_0\beta_0^2 + \beta_0^3} \quad (12)$$

Profit = total revenue generated – cost incurred by the repair man due to partial failure – cost incurred by the repair man due complete failure.

$$P_{F1} = k_0 A_{T1}(\infty) - k_1 B_{P1}(\infty) - k_2 B_{P2}(\infty) \quad (13)$$

Using the method adopted in Wang and Kuo [34], Wang and Pearn [35], Wang et al. [36] and Yen, T,-S and Wang, K.–H [37], the mathematical model of mean time to failure for Configuration 1 is derived using the relation

$$MTTF_{\cdot 1} = P(0)\left(-M_1^{-1}\right)[1,1,1,1]^T = \frac{20\alpha_0^2 + 8\alpha_0\beta_0 + \beta_0^2}{8\alpha_0^2(4\alpha_0 + \beta_0)} \quad (14)$$

$p_0(\infty) = \frac{\beta_0^3}{16\alpha_0^3 + 12\alpha_0^2\beta_0 + 4\alpha_0\beta_0^2 + \beta_0^3}$	$p_4(\infty) = \frac{4\alpha_0^2\beta_0}{16\alpha_0^3 + 12\alpha_0^2\beta_0 + 4\alpha_0\beta_0^2 + \beta_0^3}$
$p_1(\infty) = \frac{2\alpha_0\beta_0^2}{16\alpha_0^3 + 12\alpha_0^2\beta_0 + 4\alpha_0\beta_0^2 + \beta_0^3}$	$p_5(\infty) = \frac{4\alpha_0^2\beta_0}{16\alpha_0^3 + 12\alpha_0^2\beta_0 + 4\alpha_0\beta_0^2 + \beta_0^3}$
$p_2(\infty) = \frac{2\alpha_0\beta_0^2}{16\alpha_0^3 + 12\alpha_0^2\beta_0 + 4\alpha_0\beta_0^2 + \beta_0^3}$	$p_6(\infty) = \frac{8\alpha_0^3}{16\alpha_0^3 + 12\alpha_0^2\beta_0 + 4\alpha_0\beta_0^2 + \beta_0^3}$
$p_3(\infty) = \frac{4\alpha_0^2\beta_0}{16\alpha_0^3 + 12\alpha_0^2\beta_0 + 4\alpha_0\beta_0^2 + \beta_0^3}$	$p_7(\infty) = \frac{8\alpha_0^3}{16\alpha_0^3 + 12\alpha_0^2\beta_0 + 4\alpha_0\beta_0^2 + \beta_0^3}$

Table 2.
Steady state probabilities of configuration 1.

Where $P(0) = [1, 0, 0, 0]$ and

$$M_1 = \begin{pmatrix} -4\alpha_0 & 2\alpha_0 & 2\alpha_0 & 0 \\ \beta_0 & -(4\alpha_0 + \beta_0) & 0 & 2\alpha_0 \\ \beta_0 & 0 & -(4\alpha_0 + \beta_0) & 2\alpha_0 \\ 0 & \beta_0 & \beta_0 & -(4\alpha_0 + 2\beta_0) \end{pmatrix}$$

obtained by transposing Q_1 and deleting rows and columns of failure states.

4.2 Models formulation for configuration 2

Applying similar description in 4.1 above, the differential-difference equations for Configuration 2 are expressed in the form:

$$\frac{d}{dt}p(t) = Q_2 p(t) \tag{15}$$

where

$$Q_2 = \begin{pmatrix} -3\alpha_0 & \beta_0 & 0 & 0 & \beta_0 & \beta_0 & 0 & 0 & 0 & 0 & 0 & 0 & 0 \\ \alpha_0 & -y_0 & \beta_0 & 0 & 0 & 0 & \beta_0 & \beta_0 & 0 & 0 & 0 & 0 & 0 \\ 0 & \alpha_0 & -y_0 & \beta_0 & 0 & 0 & 0 & 0 & \beta_0 & \beta_0 & 0 & 0 & 0 \\ 0 & 0 & 0 & -y_0 & 0 & 0 & 0 & 0 & 0 & 0 & \beta_0 & \beta_0 & \beta_0 \\ \alpha_0 & 0 & 0 & 0 & -\beta_0 & 0 & 0 & 0 & 0 & 0 & 0 & 0 & 0 \\ \alpha_0 & 0 & 0 & 0 & 0 & -\beta_0 & 0 & 0 & 0 & 0 & 0 & 0 & 0 \\ 0 & \alpha_0 & 0 & 0 & 0 & 0 & -\beta_0 & 0 & 0 & 0 & 0 & 0 & 0 \\ 0 & \alpha_0 & 0 & 0 & 0 & 0 & 0 & -\beta_0 & 0 & 0 & 0 & 0 & 0 \\ 0 & 0 & \alpha_0 & 0 & 0 & 0 & 0 & 0 & -\beta_0 & 0 & 0 & 0 & 0 \\ 0 & 0 & \alpha_0 & 0 & 0 & 0 & 0 & 0 & 0 & -\beta_0 & 0 & 0 & 0 \\ 0 & 0 & 0 & \alpha_0 & 0 & 0 & 0 & 0 & 0 & 0 & -\beta_0 & 0 & 0 \\ 0 & 0 & 0 & \alpha_0 & 0 & 0 & 0 & 0 & 0 & 0 & 0 & -\beta_0 & 0 \\ 0 & 0 & 0 & \alpha_0 & 0 & 0 & 0 & 0 & 0 & 0 & 0 & 0 & -\beta_0 \end{pmatrix}$$

and $y_0 = (3\alpha_0 + \beta_0)$.
With initial conditions

$$P(0) = [1, 0, 0, 0, 0, 0, 0, 0, 0, 0, 0, 0, 0] \tag{16}$$

Expression for system availability, probability of partial and complete failure for Configuration 2 are given by

$$A_{T2}(\infty) = p_0(\infty) + p_1(\infty) + p_2(\infty) + p_3(\infty) \tag{17}$$

$$B_{P3}(\infty) = p_1(\infty) + p_2(\infty) + p_3(\infty) \tag{18}$$

$$B_{P4}(\infty) = p_4(\infty) + p_5(\infty) + p_6(\infty) + p_7(\infty) + \dots + p_{12}(\infty) \tag{19}$$

Setting (15) to zero to give

$$Q_2 p(\infty) = 0 \tag{20}$$

The normalizing condition for this analysis is

$$\sum_{j=0}^{12} p_j(\infty) = 1 \tag{21}$$

Combining (20) and (21) to give system of equations

$$
\begin{pmatrix}
-3\alpha_0 & \beta_0 & 0 & 0 & \beta_0 & \beta_0 & 0 & 0 & 0 & 0 & 0 & 0 & 0 \\
\alpha_0 & -y_0 & \beta_0 & 0 & 0 & 0 & \beta_0 & \beta_0 & 0 & 0 & 0 & 0 & 0 \\
0 & \alpha_0 & -y_0 & \beta_0 & 0 & 0 & 0 & 0 & \beta_0 & \beta_0 & 0 & 0 & 0 \\
0 & 0 & 0 & -y_0 & 0 & 0 & 0 & 0 & 0 & 0 & \beta_0 & \beta_0 & \beta_0 \\
\alpha_0 & 0 & 0 & 0 & -\beta_0 & 0 & 0 & 0 & 0 & 0 & 0 & 0 & 0 \\
\alpha_0 & 0 & 0 & 0 & 0 & -\beta_0 & 0 & 0 & 0 & 0 & 0 & 0 & 0 \\
0 & \alpha_0 & 0 & 0 & 0 & 0 & -\beta_0 & 0 & 0 & 0 & 0 & 0 & 0 \\
0 & \alpha_0 & 0 & 0 & 0 & 0 & 0 & -\beta_0 & 0 & 0 & 0 & 0 & 0 \\
0 & 0 & \alpha_0 & 0 & 0 & 0 & 0 & 0 & -\beta_0 & 0 & 0 & 0 & 0 \\
0 & 0 & \alpha_0 & 0 & 0 & 0 & 0 & 0 & 0 & -\beta_0 & 0 & 0 & 0 \\
0 & 0 & 0 & \alpha_0 & 0 & 0 & 0 & 0 & 0 & 0 & -\beta_0 & 0 & 0 \\
0 & 0 & 0 & \alpha_0 & 0 & 0 & 0 & 0 & 0 & 0 & 0 & -\beta_0 & 0 \\
1 & 1 & 1 & 1 & 1 & 1 & 1 & 1 & 1 & 1 & 1 & 1 & 1
\end{pmatrix}
\begin{pmatrix}
p_0(\infty) \\ p_1(\infty) \\ p_2(\infty) \\ p_3(\infty) \\ p_4(\infty) \\ p_5(\infty) \\ p_6(\infty) \\ p_7(\infty) \\ p_8(\infty) \\ p_9(\infty) \\ p_{10}(\infty) \\ p_{11}(\infty) \\ p_{12}(\infty)
\end{pmatrix}
=
\begin{pmatrix}
0 \\ 0 \\ 0 \\ 0 \\ 0 \\ 0 \\ 0 \\ 0 \\ 0 \\ 0 \\ 0 \\ 0 \\ 1
\end{pmatrix}
\tag{22}
$$

Solving the system of equations in (22) for the state probabilities $p_k(\infty), k = 0, 1, 2, \ldots, 12$, using MATLAB package, to give states probabilities in **Table 3** below.

Expressions for the system availability, probability of partial and complete failure for configuration 2 in (17) to (19) as well as profit function are now

$$A_{T2}(\infty) = \frac{\beta_0^4 + 2\alpha_0\beta_0^3 + \alpha_0^2\beta_0^2 + \alpha_0^3\beta_0}{3\alpha_0^4 + 3\alpha_0^3\beta_0 + 3\alpha_0^2\beta_0^2 + 3\alpha_0^3\beta_0 + \beta_0^4} \tag{23}$$

$$B_{P3}(\infty) = \frac{\alpha_0\beta_0^3 + \alpha_0^3\beta_0 + \alpha_0^2\beta_0^2}{3\alpha_0^4 + 3\alpha_0^3\beta_0 + 3\alpha_0^2\beta_0^2 + 3\alpha_0^3\beta_0 + \beta_0^4} \tag{24}$$

$$B_{P4}(\infty) = \frac{3\alpha_0^4 + 2\alpha_0\beta_0^3 + 2\alpha_0^3\beta_0 + 2\alpha_0^2\beta_0^2}{3\alpha_0^4 + 3\alpha_0^3\beta_0 + 3\alpha_0^2\beta_0^2 + 3\alpha_0^3\beta_0 + \beta_0^4} \tag{25}$$

$$P_{F2} = k_0 A_{T2}(\infty) - k_1 B_{P3}(\infty) - k_2 B_{P4}(\infty) \tag{26}$$

$$p_0(\infty) = \frac{\beta_0^4}{3\alpha_0^4 + 3\alpha_0^3\beta_0 + 3\alpha_0^2\beta_0^2 + 3\alpha_0^3\beta_0 + \beta_0^4}$$

$$p_1(\infty) = \frac{\alpha_0\beta_0^3}{3\alpha_0^4 + 3\alpha_0^3\beta_0 + 3\alpha_0^2\beta_0^2 + 3\alpha_0^3\beta_0 + \beta_0^4}$$

$$p_2(\infty) = \frac{\alpha_0^2\beta_0^2}{3\alpha_0^4 + 3\alpha_0^3\beta_0 + 3\alpha_0^2\beta_0^2 + 3\alpha_0^3\beta_0 + \beta_0^4}$$

$$p_3(\infty) = \frac{\alpha_0^3\beta_0}{3\alpha_0^4 + 3\alpha_0^3\beta_0 + 3\alpha_0^2\beta_0^2 + 3\alpha_0^3\beta_0 + \beta_0^4}$$

$$p_4(\infty) = \frac{\alpha_0^2\beta_0}{3\alpha_0^4 + 3\alpha_0^3\beta_0 + 3\alpha_0^2\beta_0^2 + 3\alpha_0^3\beta_0 + \beta_0^4}$$

$$p_5(\infty) = \frac{\alpha_0^3\beta_0}{3\alpha_0^4 + 3\alpha_0^3\beta_0 + 3\alpha_0^2\beta_0^2 + 3\alpha_0^3\beta_0 + \beta_0^4}$$

$$p_6(\infty) = \frac{\alpha_0^2\beta_0^2}{3\alpha_0^4 + 3\alpha_0^3\beta_0 + 3\alpha_0^2\beta_0^2 + 3\alpha_0^3\beta_0 + \beta_0^4}$$

$$p_7(\infty) = \frac{\alpha_0^2\beta_0^2}{3\alpha_0^4 + 3\alpha_0^3\beta_0 + 3\alpha_0^2\beta_0^2 + 3\alpha_0^3\beta_0 + \beta_0^4}$$

$$p_8(\infty) = \frac{\alpha_0^3\beta_0}{3\alpha_0^4 + 3\alpha_0^3\beta_0 + 3\alpha_0^2\beta_0^2 + 3\alpha_0^3\beta_0 + \beta_0^4}$$

$$p_9(\infty) = \frac{\alpha_0^3\beta_0}{3\alpha_0^4 + 3\alpha_0^3\beta_0 + 3\alpha_0^2\beta_0^2 + 3\alpha_0^3\beta_0 + \beta_0^4}$$

$$p_{10}(\infty) = \frac{\alpha_0^4}{3\alpha_0^4 + 3\alpha_0^3\beta_0 + 3\alpha_0^2\beta_0^2 + 3\alpha_0^3\beta_0 + \beta_0^4}$$

$$p_{11}(\infty) = \frac{\alpha_0^4}{3\alpha_0^4 + 3\alpha_0^3\beta_0 + 3\alpha_0^2\beta_0^2 + 3\alpha_0^3\beta_0 + \beta_0^4}$$

$$p_{12}(\infty) = \frac{\alpha_0^4}{3\alpha_0^4 + 3\alpha_0^3\beta_0 + 3\alpha_0^2\beta_0^2 + 3\alpha_0^3\beta_0 + \beta_0^4}$$

Table 3.
Steady state probabilities of configuration 2.

Mathematical model of mean time to failure for Configuration 2 is derived using the relation

$$MTTF_2 = P(0)\left(-M_2^{-1}\right)[1,1,1,1]^T = \frac{40\alpha_0^3 + 27\alpha_0^2\beta_0 + 8\alpha_0\beta_0^2 + \beta_0^3}{\alpha_0\left(81\alpha_0^3 + 54\alpha_0^2\beta_0 + 16\alpha_0\beta_0^2 + 2\beta_0^3\right)} \quad (27)$$

Where $P(0) = [1,0,0,0]$ and

$$M_2 = \begin{pmatrix} -3\alpha_0 & \alpha_0 & 0 & 0 \\ \beta_0 & -(3\alpha_0 + \beta_0) & \alpha_0 & 0 \\ 0 & \beta_0 & -(3\alpha_0 + \beta_0) & \alpha_0 \\ 0 & 0 & \beta_0 & -(3\alpha_0 + \beta_0) \end{pmatrix}$$

M_2 is obtained from Q_2 using similar argument above.

4.3 Models formulation for configuration 3

Following similar argument in 4.1 above, the differential-difference equations obtained for Configuration 3 are expressed in the form:

$$\frac{d}{dt}p(t) = Q_3 p(t) \quad (28)$$

where

$$Q_3 = \begin{pmatrix} -2\alpha_0 & \beta_0 & \beta_0 & 0 & 0 & 0 & 0 & 0 & 0 & 0 & 0 \\ \alpha_0 & -y_1 & 0 & \beta_0 & \beta_0 & 0 & 0 & 0 & 0 & 0 & 0 \\ \alpha_0 & 0 & -y_1 & 0 & 0 & 0 & 0 & 0 & 0 & \beta_0 & \beta_0 \\ 0 & \alpha_0 & 0 & -y_1 & 0 & \beta_0 & \beta_9 & 0 & 0 & 0 & 0 \\ 0 & \alpha_0 & 0 & 0 & -y_1 & 0 & 0 & \beta_0 & \beta_0 & 0 & 0 \\ 0 & 0 & 0 & \alpha_0 & 0 & -\beta_0 & 0 & 0 & 0 & 0 & 0 \\ 0 & 0 & 0 & \alpha_0 & 0 & 0 & -\beta_0 & 0 & 0 & 0 & 0 \\ 0 & 0 & 0 & 0 & \alpha_0 & 0 & 0 & -\beta_0 & 0 & 0 & 0 \\ 0 & 0 & 0 & 0 & \alpha_0 & 0 & 0 & 0 & -\beta_0 & 0 & 0 \\ 0 & 0 & \alpha_0 & 0 & 0 & 0 & 0 & 0 & 0 & -\beta_0 & 0 \\ 0 & 0 & \alpha_0 & 0 & 0 & 0 & 0 & 0 & 0 & 0 & -\beta_0 \end{pmatrix}$$

Where $y_1 = (2\alpha_0 + \beta_0)$.
With initial conditions

$$P(0) = [1,0,0,0,0,0,0,0,0,0,0] \quad (29)$$

Expression for system availability, probability of partial and complete failure for Configuration 3 are given by

$$A_{T3}(\infty) = p_0(\infty) + p_1(\infty) + p_2(\infty) + p_3(\infty) + p_4(\infty) \quad (30)$$

$$B_{P5}(\infty) = p_1(\infty) + p_2(\infty) + p_3(\infty) + p_4(\infty) \quad (31)$$

$$B_{P6}(\infty) = p_5(\infty) + p_6(\infty) + p_7(\infty) + p_8(\infty) + p_9(\infty) + p_{10}(\infty) \quad (32)$$

Setting (28) to zero to give

$$Q_3 p(\infty) = 0 \tag{33}$$

The normalizing condition for this analysis is

$$\sum_{j=0}^{10} p_j(\infty) = 1 \tag{34}$$

Combining (33) and (34) to give system of equations

$$
\begin{pmatrix}
-2\alpha_0 & \beta_0 & \beta_0 & 0 & 0 & 0 & 0 & 0 & 0 & 0 & 0 \\
\alpha_0 & -y_1 & 0 & \beta_0 & \beta_0 & 0 & 0 & 0 & 0 & 0 & 0 \\
\alpha_0 & 0 & -y_1 & 0 & 0 & 0 & 0 & 0 & 0 & \beta_0 & \beta_0 \\
0 & \alpha_- & 0 & -y_1 & 0 & \beta_0 & \beta_9 & 0 & 0 & 0 & 0 \\
0 & \alpha_0 & 0 & 0 & -y_1 & 0 & 0 & \beta_0 & \beta_0 & 0 & 0 \\
0 & 0 & 0 & \alpha_0 & 0 & -\beta_0 & 0 & 0 & 0 & 0 & 0 \\
0 & 0 & 0 & \alpha_0 & 0 & 0 & -\beta_0 & 0 & 0 & 0 & 0 \\
0 & 0 & 0 & 0 & \alpha_0 & 0 & 0 & -\beta_0 & 0 & 0 & 0 \\
0 & 0 & 0 & 0 & \alpha_0 & 0 & 0 & 0 & -\beta_0 & 0 & 0 \\
0 & 0 & \alpha_0 & 0 & 0 & 0 & 0 & 0 & 0 & -\beta_0 & 0 \\
1 & 1 & 1 & 1 & 1 & 1 & 1 & 1 & 1 & 1 & 1
\end{pmatrix}
\begin{pmatrix}
p_0(\infty) \\
p_1(\infty) \\
p_2(\infty) \\
p_3(\infty) \\
p_4(\infty) \\
p_5(\infty) \\
p_6(\infty) \\
p_7(\infty) \\
p_8(\infty) \\
p_9(\infty) \\
p_{10}(\infty)
\end{pmatrix}
=
\begin{pmatrix}
0 \\ 0 \\ 0 \\ 0 \\ 0 \\ 0 \\ 0 \\ 0 \\ 0 \\ 0 \\ 1
\end{pmatrix}
\tag{35}
$$

Solving the system of equations in (35) for the state probabilities $p_k(\infty), k = 0, 1, 2, \ldots, 10$, presented in **Table 4** below.

Expressions for the system availability, probability of partial and complete failure for configuration 3 in (30) to (32) as well as profit function are now

$$A_{T3}(\infty) = \frac{\beta_0^3 + 2\alpha_0\beta_0^2 + 2\alpha_0^2\beta_0}{4\alpha_0^3 + 4\alpha_0^2\beta_0 + 2\alpha_0\beta_0^2 + \beta_0^3} \tag{36}$$

$$B_{P5}(\infty) = \frac{2\alpha_0\beta_0(\alpha_0 + \beta_0)}{4\alpha_0^3 + 4\alpha_0^2\beta_0 + 2\alpha_0\beta_0^2 + \beta_0^3} \tag{37}$$

$$B_{P6}(\infty) = \frac{2\alpha_0^2(2\alpha_0 + \beta_0)}{4\alpha_0^3 + 4\alpha_0^2\beta_0 + 2\alpha_0\beta_0^2 + \beta_0^3} \tag{38}$$

$p_0(\infty) = \frac{\beta_0^3}{4\alpha_0^3 + 4\alpha_0^2\beta_0 + 2\alpha_0\beta_0^2 + \beta_0^3}$

$p_6(\infty) = \frac{\alpha_0^3}{4\alpha_0^3 + 4\alpha_0^2\beta_0 + 2\alpha_0\beta_0^2 + \beta_0^3}$

$p_1(\infty) = \frac{\alpha_0\beta_0^2}{4\alpha_0^3 + 4\alpha_0^2\beta_0 + 2\alpha_0\beta_0^2 + \beta_0^3}$

$p_7(\infty) = \frac{\alpha_0^3}{4\alpha_0^3 + 4\alpha_0^2\beta_0 + 2\alpha_0\beta_0^2 + \beta_0^3}$

$p_2(\infty) = \frac{\alpha_0\beta_0^2}{4\alpha_0^3 + 4\alpha_0^2\beta_0 + 2\alpha_0\beta_0^2 + \beta_0^3}$

$p_8(\infty) = \frac{\alpha_0^3}{4\alpha_0^3 + 4\alpha_0^2\beta_0 + 2\alpha_0\beta_0^2 + \beta_0^3}$

$p_3(\infty) = \frac{\alpha_0^2\beta_0}{4\alpha_0^3 + 4\alpha_0^2\beta_0 + 2\alpha_0\beta_0^2 + \beta_0^3}$

$p_9(\infty) = \frac{\alpha_0^2\beta_0}{4\alpha_0^3 + 4\alpha_0^2\beta_0 + 2\alpha_0\beta_0^2 + \beta_0^3}$

$p_4(\infty) = \frac{\alpha_0^2\beta_0}{4\alpha_0^3 + 4\alpha_0^2\beta_0 + 2\alpha_0\beta_0^2 + \beta_0^3}$

$p_{10}(\infty) = \frac{\alpha_0^2\beta_0}{4\alpha_0^3 + 4\alpha_0^2\beta_0 + 2\alpha_0\beta_0^2 + \beta_0^3}$

$p_5(\infty) = \frac{\alpha_0^3}{4\alpha_0^3 + 4\alpha_0^2\beta_0 + 2\alpha_0\beta_0^2 + \beta_0^3}$

Table 4.
Steady state probabilities of configuration 3.

$$P_{F3} = k_0 A_{T3}(\infty) - k_1 B_{P5}(\infty) - k_2 B_{P6}(\infty) \tag{39}$$

Mathematical model of mean time to failure for Configuration 3 is derived using the relation

$$MTTF_3 = P(0)\left(-M_3^{-1}\right)[1,1,1,1,1]^T = \frac{20\alpha_0^3 + 16\alpha_0^2\beta_0 + 6\alpha_0\beta_0^2 + \beta_0^3}{2\alpha_0^2\left(8\alpha_0^2 + 4\alpha_0\beta_0 + \beta_0^2\right)} \tag{40}$$

Where $P(0) = [1,0,0,0,0]$

$$M_3 = \begin{pmatrix} -2\alpha_0 & \alpha_0 & \alpha_0 & 0 & 0 \\ \beta_0 & -(2\alpha_0 + \beta_0) & 0 & \alpha_0 & \alpha_0 \\ \beta_0 & 0 & -(2\alpha_0 + \beta_0) & 0 & 0 \\ 0 & \beta_0 & 0 & -(2\alpha_0 + \beta_0) & 0 \\ 0 & \beta_0 & 0 & 0 & -(2\alpha_0 + \beta_0) \end{pmatrix}$$

M_3 is obtained from Q_3 using similar argument above.

5. Results and discussion

5.1 Analytical comparison

In this section, the configurations are compared analytically in terms of their availability and mean time to failure to determine the optimal configuration by taking the difference between mean time to failure (MTTF) and availability for the configurations using MAPLE software package.

$$MTTF_3 - MTTF_1 = \frac{16\alpha_0^4 + 192\alpha_0^3\beta_0 + 100\alpha_0^2\beta_0^2 + 28\alpha_0\beta_0^3 + 3\beta_0^4}{8\alpha_0^2(4\alpha_0 + \beta_0)(8\alpha_0^2 + 4\alpha_0\beta_0 + \beta_0^2)} \tag{41}$$

$$\Rightarrow MTTF_3 > MTTF_1 \qquad \forall \alpha_0, \beta_0 > 0$$

$$MTTF_3 - MTTF_2 = \frac{980\alpha_0^6 + 1624\alpha_0^5\beta_0 + 1246\alpha_0^4\beta_0^2 + 567\alpha_0^3\beta_0^3 + 158\alpha_0^2\beta_0^4 + 26\alpha_0\beta_0^5 + 2\beta_0^6}{2\alpha_0^2(8\alpha_0^2 + 4\alpha_0\beta_0 + \beta_0^2)(81\alpha_0^3 + 54\alpha_0^2\beta_0 + 16\alpha_0\beta_0^2 + 2\beta_0^3)} \tag{42}$$

$$\Rightarrow MTTF_3 > MTTF_2 \qquad \forall \alpha_0, \beta_0 > 0$$

$$MTTF_1 - MTTF_2 = \frac{340\alpha_0^5 + 544\alpha_0^4\beta_0 + 361\alpha_0^3\beta_0^2 + 126\alpha_0^2\beta_0^3 + 24\alpha_0\beta_0^4 + 2\beta_0^5}{8\alpha_0^2(4\alpha_0 + \beta_0)(81\alpha_0^3 + 54\alpha_0^2\beta_0 + 16\alpha_0\beta_0^2 + 2\beta_0^3)} \tag{43}$$

$$\Rightarrow MTTF_1 > MTTF_2 \qquad \forall \alpha_0, \beta_0 > 0$$

Using mean time to failure models of configurations, it is clear from (41)–(43) that

$$MTTF_3 > MTTF_1 > MTTF_2$$

$$A_{T3}(\infty) - A_{T1}(\infty) = \frac{2\alpha_0^2\beta_0\left(4\alpha_0^2 + 4\alpha_0\beta_0 + 3\beta_0^2\right)}{\left(16\alpha_0^3 + 12\alpha_0^2\beta_0 + 4\alpha_0\beta_0^2 + \beta_0^3\right)\left(4\alpha_0^3 + 4\alpha_0^2\beta_0 + 2\alpha_0\beta_0^2 + \beta_0^3\right)} \tag{44}$$

$$\Rightarrow A_{T3} > A_{T1} \qquad \forall \alpha_0, \beta_0 > 0$$

$$A_{T3}(\infty) - A_{T2}(\infty) = \frac{\alpha_0 \beta_0 \left(2\alpha_0^5 + 4\alpha_0^4 \beta_0 + \alpha_0^3 \beta_0^2 + 2\alpha_0 \beta_0^4 + \beta_0^5\right)}{\left(4\alpha_0^3 + 4\alpha_0^2 \beta_0 + 2\alpha_0 \beta_0^2 + \beta_0^3\right)\left(3\alpha_0^4 + 3\alpha_0^3 \beta_0 + 3\alpha_0^2 \beta_0^2 + 3\alpha_0 \beta_0^3 + \beta_0^4\right)} \tag{45}$$

$$\Rightarrow A_{T3} > A_{T2} \qquad \forall \alpha_0, \beta_0 > 0$$

$$A_{T2}(\infty) - A_{T1}(\infty) = \frac{\alpha_0 \beta_0 \left(2\alpha_0^4 + \alpha_0^3 \beta_0 + 10\alpha_0^2 \beta_0^2 + 4\alpha_0 \beta_0^3 - \beta_0^4\right)}{\left(3\alpha_0^4 + 3\alpha_0^3 \beta_0 + 3\alpha_0^2 \beta_0^2 + 3\alpha_0 \beta_0^3 + \beta_0^4\right)\left(16\alpha_0^3 + 12\alpha_0^2 \beta_0 + 4\alpha_0 \beta_0^2 + \beta_0^3\right)} \tag{46}$$

$$\Rightarrow A_{T2} > A_{T1} \qquad \forall \alpha_0, \beta_0 > 0$$

Using availability models of configurations, it is clear from (44)–(46) that

$$A_{T3}(\infty) > A_{T2}(\infty) > A_{T1}(\infty)$$

5.2 Comparison based on ranking of the configurations

Tables 5 and 6 depict the ranking of configuration base on their availability and mean time to failure. It clear from **Table 5** that configuration 3 is the optimal configuration whenever $0 \leq \beta_0 \leq 1$. Thus, verifying the above analytical claim that $A_{T3}(\infty) > A_{T2}(\infty) > A_{T1}(\infty)$ and $MTTF_3 > MTTF_1 > MTTF_2$ $\alpha_0, \beta_0 > 0$.

5.3 Comparison based on availability, profit and mean time to failure

In this section, $\beta_0 = 0.8$, $k_0 = 50,000,000$, $k_1 = 1250$ and $k_2 = 2150$ are fixed and vary α_0 from 0.1 to 1 in **Figures 4–6** and $\alpha_0 = 0.018$, $k_0 = 50,000,000$,

Case	Parameter Range	Result		Constant parameters
1	$0 < \alpha_0 < 0.2$	$A_{V3}(\infty) > A_{V2}(\infty) > A_{V1}(\infty)$	$MTTF_3 > MTTF_1 > MTTF_2$	$\beta_0 = 0.6$
	$0.2 < \alpha_0 < 0.4$	$A_{V3}(\infty) > A_{V2}(\infty) > A_{V1}(\infty)$	$MTTF_3 > MTTF_1 > MTTF_2$	
	$0.4 < \alpha_0 < 0.6$	$A_{V3}(\infty) > A_{V2}(\infty) > A_{V1}(\infty)$	$MTTF_3 > MTTF_1 > MTTF_2$	
	$0.6 < \alpha_0 < 0.8$	$A_{V3}(\infty) > A_{V2}(\infty) > A_{V1}(\infty)$	$MTTF_3 > MTTF_1 > MTTF_2$	
	$0.8 < \alpha_0 < 1.0$	$A_{V3}(\infty) > A_{V2}(\infty) > A_{V1}(\infty)$	$MTTF_3 > MTTF_1 > MTTF_2$	

Table 5.
Ranking of configurations based on their availability and MTTF for $\alpha_o \in [0, 1]$.

Case	Parameter Range	Result		Constant parameters
2	$0 < \beta_0 < 0.2$	$A_{V3}(\infty) > A_{V2}(\infty) > A_{V1}(\infty)$	$MTTF_3 > MTTF_1 > MTTF_2$	$\alpha_0 = 0.2$
	$0.2 < \beta_0 < 0.4$	$A_{V3}(\infty) > A_{V2}(\infty) > A_{V1}(\infty)$	$MTTF_3 > MTTF_1 > MTTF_2$	
	$0.4 < \beta_0 < 0.6$	$A_{V3}(\infty) > A_{V2}(\infty) > A_{V1}(\infty)$	$MTTF_3 > MTTF_1 > MTTF_2$	
	$0.6 < \beta_0 < 0.8$	$A_{V3}(\infty) > A_{V2}(\infty) > A_{V1}(\infty)$	$MTTF_3 > MTTF_1 > MTTF_2$	
	$0.8 < \beta_0 < 1.0$	$A_{V3}(\infty) > A_{V2}(\infty) > A_{V1}(\infty)$	$MTTF_3 > MTTF_1 > MTTF_2$	

Table 6.
Ranking of configurations based on their availability and MTTF for $\beta_o \in [0, 1]$.

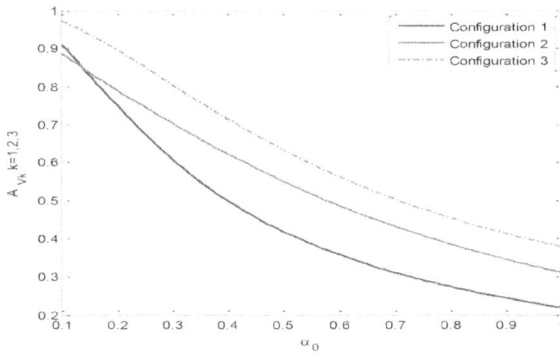

Figure 4.
Availability against α_0.

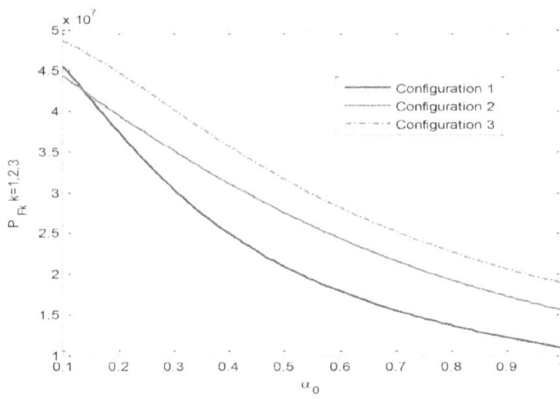

Figure 5.
Profit against α_0.

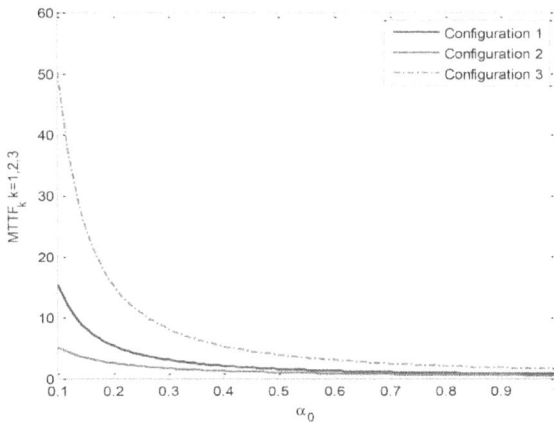

Figure 6.
MTTF against α_0.

$k_1 = 1250$ and $k_2 = 2150$ are fixed and vary β_0 from 0.1 to 1 in **Figures 7–9** and obtained the following results.

Simulations in **Figures 4–6** compare the steady state availability, profit and MTTF with respect to α_0 for all the three configurations considered. From these figures, availability, profit and MTTF decreases as α_0 increases for any configuration. Furthermore, Configuration 3 seems to be most effective and reliable configuration among all the three configurations. From these figures, it is clear that Configuration 3 produces more availability, profit and MTTF than the other configurations. Thus, Configuration 3 is the optimal configuration in this study. On the other hand, simulations in **Figures 7–9** compare the steady state availability, profit and MTTF with respect to β_0 for all the three configurations. It is evident from these figures that availability, profit and MTTF increases as β_0 increases for any

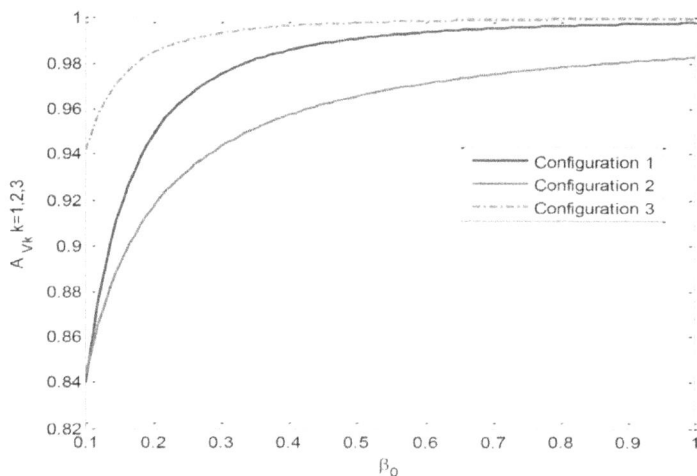

Figure 7.
Availability against β_0.

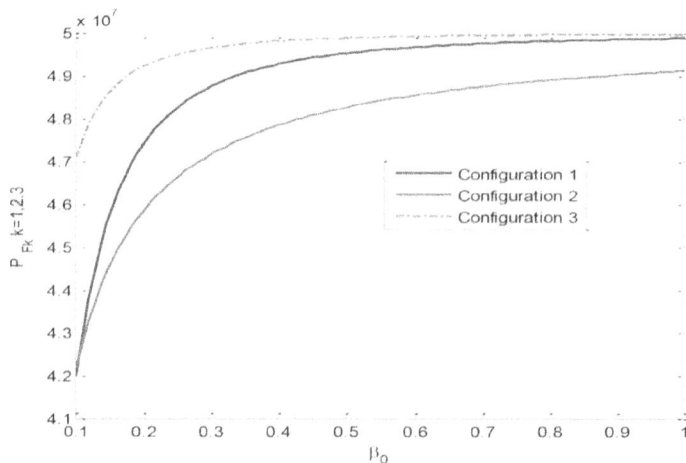

Figure 8.
Profit against β_0.

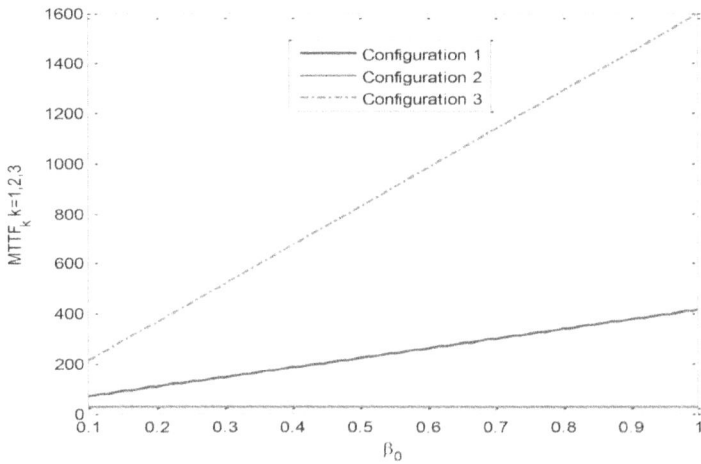

Figure 9.
MTTF against β_o.

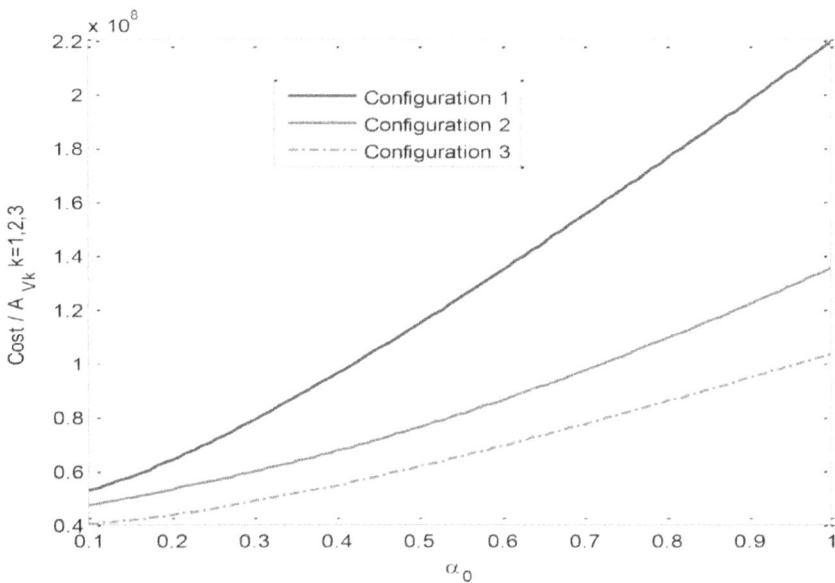

Figure 10.
C_k/A_{Vk} against α_o.

configuration. Similar to **Figures 4–6**, Configuration 3 seems to be most effective and reliable configuration among all the three configurations and hence is the optimal configuration.

5.4 Comparison based on cost benefit

In this section, the configurations are compared based on their C_k/A_{vk} and $C_k/MTTF_k$ using MATLAB software. The following parameters values are used for the purpose of analysis:

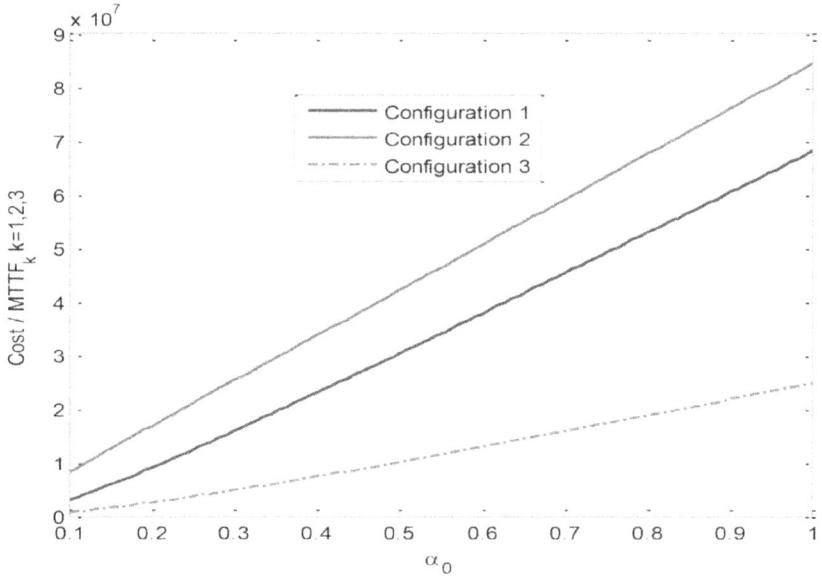

Figure 11.
$C_k/MTTF_k$ against α_0.

$\beta_0 = 0.8$, $C_1 = 48,000,000$, $C_2 = 42,000,000$, $C_3 = 39,000,000$ (Yen, T,-S and Wang, K.–H [37]) are fixed and vary α_0 between 0.1 and 1 in **Figures 10** and **11**.

$\alpha_0 = 0.018$,, $C_1 = 42,000,000$, $C_1 = 39,000,000$ and vary β_0 between 0.1 to 1 in **Figures 12** and **13** and obtained the following results:

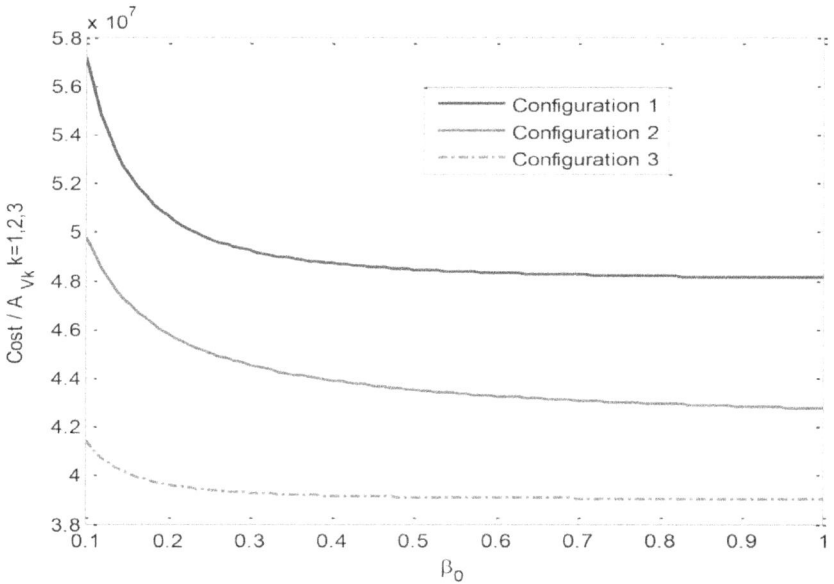

Figure 12.
C_k/A_{Vk} against β_0.

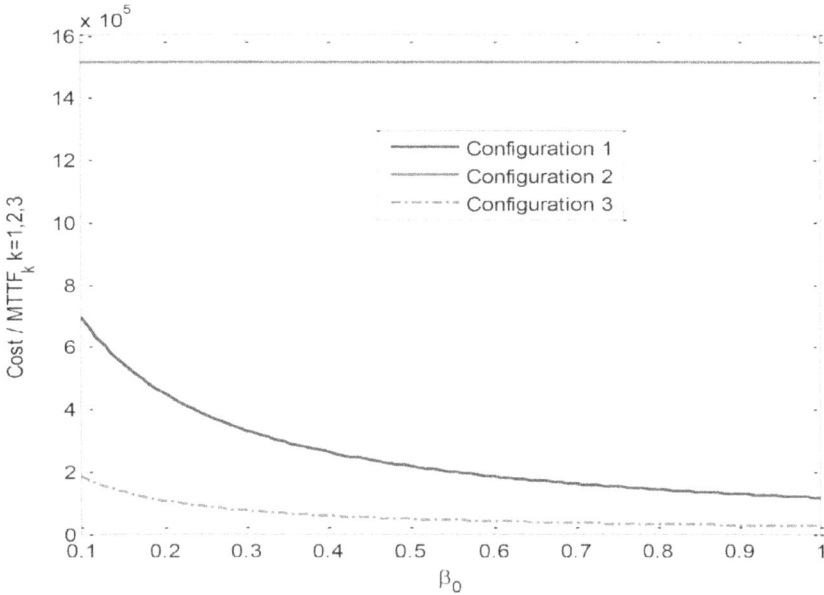

Figure 13.
$C_k/MTTF_k$ against β_o.

Figures 10 and **11** depict the results of C_k/A_{vk} and $C_k/MTTF_k$ for each configuration i ($i = 1, 2, 3$) with respect to α_0. From these figures, it is evident that C_k/A_{vk} and $C_k/MTTF_k$ increase as α_0 increases for each configuration. It is evident from these Figures that the optimal configuration using both C_k/A_{vk} and $C_k/MTTF_k$ is Configuration 3.

From **Figures 12** and **13**, it is clear that C_k/A_{vk} and $C_k/MTTF_k$ decrease as β_0 increases. It is clear from these Figures that the optimal configuration using both C_k/A_{vk} and $C_k/MTTF_k$ is again Configuration 3. Configurations 1 and 2 tend to have more C_k/A_{vk} and $C_k/MTTF_k$ that Configuration 3. From the result presented in this study, it is clear that the survival of manufacturing and industrial systems depends upon its design and reliability characteristics. Through the system design and reliability characteristics, management can tend to realize whether such systems operate at minimum cost of maintenance, quality of the product, production output as well as revenue mobilization.

5.5 Sensitivity analysis

Sensitivity analysis presented in **Tables 7** and **8** depict the change in availability, MTTF and profit of the three configurations with respect to failure rate α_0 for different values of β_0. It is clear from these tables that availability, MTTF and profit of the three configurations decreases as α_0 increase. Availability, MTTF and profit tend to be higher for the three configurations for whenever $\beta_0 = 0.9$. This sensitivity analysis implies that maintenance action such as inspection, preventive maintenance, etc. should be invoke to reduce the occurrence of failure in order to attain maximum value of availability, MTTF and profit. From these tables it is evident that Configuration 3 has higher values of availability, MTTF and profit than configurations 1 and 2 for different values of β_0. On the other hand, Sensitivity analysis in depicted in **Tables 9** and **10**, displayed the variation of availability, MTTF and

	α_0	$A_{V1}(\infty)$	$A_{V2}(\infty)$	$A_{V3}(\infty)$	$MTTF_1$	$MTTF_2$	$MTTF_3$
$\beta_0 = 0.3$	0.1	0.6522	0.7313	0.8361	9.4643	4.9887	25.6897
	0.2	0.3962	0.5287	0.6084	3.8920	2.4878	9.3654
	0.3	0.2727	0.3846	0.4545	2.4167	1.6558	5.5128
	0.4	0.2050	0.2902	0.3565	1.7475	1.2405	3.8699
	0.5	0.1632	0.2286	0.2912	1.3674	0.9916	2.9717
$\beta_0 = 0.6$	0.1	0.8571	0.8538	0.9494	13.0000	4.9965	40.2941
	0.2	0.6522	0.7313	0.8361	4.7321	2.4943	12.8448
	0.3	0.5000	0.6230	0.7143	2.7778	1.6605	7.0000
	0.4	0.3962	0.5287	0.6084	1.9460	1.2439	4.6827
	0.5	0.3243	0.4494	0.5227	1.4923	0.9942	3.4809
$\beta_0 = 0.9$	0.1	0.9252	0.8989	0.9764	16.6346	4.9985	55.1600
	0.2	0.7852	0.8112	0.9154	5.6066	2.4970	16.4662
	0.3	0.6522	0.7313	0.8361	3.1548	1.6629	8.5632
	0.4	0.5445	0.6576	0.7537	2.1531	1.2459	5.5391
	0.5	0.4609	0.5899	0.6768	1.6224	0.9959	4.0169

Table 7.
Availability and MTTF sensitivity as function of α_0 for different values of β_0.

α_0	Profit $* 10^7$								
	$\beta_0 = 0.3$			$\beta_0 = 0.6$			$\beta_0 = 0.9$		
	P_{F1}	P_{F2}	P_{F3}	P_{F1}	P_{F2}	P_{F3}	P_{F1}	P_{F2}	P_{F3}
0	5.0000	5.0000	5.0000	5.0000	5.0000	5.0000	5.0000	5.0000	5.0000
0.1	3.0666	3.5304	4.0420	4.1659	4.1968	4.6934	4.5546	4.4432	4.8555
0.2	1.8056	2.4578	2.8393	3.0666	3.5304	4.0420	3.7692	3.9643	4.4934
0.3	1.2298	1.7423	2.0850	2.3044	2.9493	3.3837	3.0666	3.5304	4.0420
0.4	0.9205	1.2982	1.6218	1.8056	2.4578	2.8393	2.5232	3.1335	3.5928
0.5	0.7316	1.0172	1.3190	1.4682	2.0586	2.4143	2.1147	2.7751	3.1881
0.6	0.6054	0.8304	1.1085	1.2298	1.7423	2.0850	1.8056	2.4578	2.8393
0.7	0.5156	0.6995	0.9547	1.0542	1.4939	1.8270	1.5673	2.1818	2.5441
0.8	0.4486	0.6035	0.8378	0.9205	1.2982	1.6218	1.3800	1.9446	2.2952
0.9	0.3968	0.5305	0.7461	0.8157	1.1426	1.4556	1.2298	1.7423	2.0850

Table 8.
Profit sensitivity as function of α_0 for different values of β_0.

profit with respect to β_0 for different values of α_0. It is evident from the tables that availability, MTTF and profit increases as β_0 increases for different values of α_0. Increase in the values of α_0 decrease the availability, MTTF and profit as shown in the tables. This sensitivity analysis suggest that perfect repair, preventive maintenance, inspection should be invoke at early failure to restore the system to its position as good as new. It is also clear from these tables that Configuration III has higher values of availability, MTTF and profit than configurations 1 and 2.

	α_0	$A_{V1}(\infty)$	$A_{V2}(\infty)$	$A_{V3}(\infty)$	$MTTF_1$	$MTTF_2$	$MTTF_3$
$\alpha_0 = 0.012$	0.1	0.9150	0.8915	0.9726	131.8506	41.6518	432.21
	0.2	0.9749	0.9432	0.9929	217.2939	41.6638	778.30
	0.3	0.9883	0.9615	0.9968	303.5201	41.6657	1125.2
	0.4	0.9933	0.9708	0.9982	390.0050	41.6662	1472.4
	0.5	0.9956	0.9765	0.9989	476.6069	41.6664	1819.5
$\alpha_0 = 0.015$	0.1	0.8784	0.8671	0.9583	92.0139	33.3142	290.60
	0.2	0.9623	0.9299	0.9890	146.3675	33.3294	511.70
	0.3	0.9821	0.9523	0.9950	201.3889	33.3319	733.60
	0.4	0.9896	0.9638	0.9972	256.6425	33.3327	955.70
	0.5	0.9933	0.9708	0.9982	312.0040	33.3330	1177.90
$\alpha_0 = 0.018$	0.1	0.8399	0.8434	0.9418	69.2650	27.7550	211.6955
	0.2	0.9479	0.9168	0.9843	106.7765	27.7726	364.8292
	0.3	0.9749	0.9432	0.9929	144.8626	27.7759	518.8338
	0.4	0.9854	0.9569	0.9960	183.1581	27.7769	673.0276
	0.5	0.9904	0.9652	0.9974	221.2852	27.7773	827.2852

Table 9.
Availability and MTTF sensitivity as function of β_o for different values of α_o.

β_0	Profit $* 10^7$								
	$\alpha_0 = 0.012$			$\alpha_0 = 0.015$			$\alpha_0 = 0.018$		
	P_{F1}	P_{F2}	P_{F3}	P_{F1}	P_{F2}	P_{F3}	P_{F1}	P_{F2}	P_{F3}
0	−0.0002	−0.0002	−0.0002	−0.0002	−0.0002	−0.0002	−0.0002	−0.0002	−0.0002
0.1	4.6436	4.5075	4.8881	4.4850	4.3959	4.8289	4.3156	4.2875	4.7597
0.2	4.8969	4.7431	4.9711	4.8444	4.6825	4.9552	4.7839	4.6230	4.9359
0.3	4.9521	4.8260	4.9871	4.9268	4.7842	4.9799	4.8969	4.7431	4.9711
0.4	4.9725	4.8684	4.9927	4.9577	4.8366	4.9887	4.9400	4.8051	4.9837
0.5	4.9822	4.8942	4.9953	4.9725	4.8684	4.9927	4.9609	4.8429	4.9895
0.6	4.9875	4.9116	4.9968	4.9807	4.8899	4.9949	4.9725	4.8684	4.9927
0.7	4.9908	4.9240	4.9976	4.9857	4.9054	4.9963	4.9796	4.8868	4.9946
0.8	4.9929	4.9334	4.9982	4.9890	4.9170	4.9972	4.9843	4.9007	4.9959
0.9	4.9944	4.9407	4.9986	4.9913	4.9261	4.9977	4.9875	4.9116	4.9968

Table 10.
Profit sensitivity as function of β_o for different values of α_o.

6. Conclusion

In this paper, three different standby serial systems each supplying 60 MW are considered. The expressions for the reliability characteristics such as system availability, busy period of repairman due to partial and complete failure as well as profit functions for all the configurations have been obtained and validated by performing analytical and numerical experiments. Analysis of the effect of various system parameters on mean time to failure, profit function and availability was performed.

These are the main contributions of this study. On the basis of the numerical results obtained in Figures and Tables for a particular case, it is evident that the optimal system configuration is configuration 3. This is supported from analytical comparison presented in terms of the availability and mean time to failure models obtained in which configuration III is the optimal configuration for all $\alpha_0, \beta_0 > 0$ contrary to some studies where the optimality among the system configuration is not uniform as it depends on some system parameters. The contributions of this paper are as follows.

i. Failure is categorized into partial and complete failure

ii. Analytical comparison between the configuration in terms of their availability and mean time to failure is performed

iii. Optimal configuration in analytical comparison agrees with that of numerical comparison

iv. Optimal configuration is unique for all parameter values

Conflict of interest

There are no conflicts of interest to this chapter.

Additional classification

AMS (2010) subject classification: 90B25

Author details

Ibrahim Yusuf[1]* and Ismail Muhammad Musa[2]

1 Department of Mathematical Sciences, Bayero University, Kano, Nigeria

2 Department of Mathematics and Computer Science, Alqalam University, Katsina, Nigeria

*Address all correspondence to: iyusuf.mth@buk.edu.ng

IntechOpen

References

[1] Singh VV, Gulati J, Rawal DK, Goel CK. Performance analysis of complex system in series configuration under different failure and repair discipline using copula. International Journal of Reliability Quality and Safety Engineering. 2016a;*23*(2)

[2] Singh VV, Gulati J, Rawal DK, Goel CK. Performance analysis of complex system in series configuration under different failure and repair discipline using copula. International Journal of Reliability Quality and Safety Engineering. 2016b;*23*(2)

[3] Lado AI, Singh VV. Cost assessment of complex repairable system consisting of two subsystems in the series configuration using Gumbel–Hougaard family copula. Int. J. Qual. Rel. Manage. 2019;*36*(10):1683-1698

[4] Yusuf I. Reliability Modeling of a Parallel System with a Supporting Device and Two Types Preventive Maintenance. International journal of operational Research. 2016;*25*(3): 269-287

[5] Singh VV, Ayagi HI. Stochastic analysis of a complex system under preemptive resume repair policy using Gumbel-Hougaard family of copula. International Journal of Mathematics in Operational Research. 2018;*12*(2): 273-292

[6] Niwas R, Garg H. An approach for analyzing the reliability and profit of an industrial system based on the cost free warranty policy. Journal of the Brazilian Society of Mechanical Sciences and Engineering. 2018;*40*:265

[7] Monika Gahlot, V. V. Singh, H. Ismail Ayagi & C.K. Goel. (2018) Performance assessment of repairable system in series configuration under different types of failure and repair policies using Copula Linguistics",

International Journal of Reliability and Safety. Vol.12 Issue 4 pp.348-374.

[8] Gahlot M, Singh VV, Ayagi H, Goel CK. Performance assessment of repairable system in series configuration under different types of failure and repair policies using copula linguistics. Int. J. Rel. Safe. 2018;*12*(4):348-374

[9] Singh, V.V and Singh, N. P. Performance analysis of three unit redundant system with switch and human failure. Math. Eng., Sci. Aerosp. 2015;*6*(2):295-308

[10] Saini M, Kumar A. Performance analysis of evaporation system in sugar industry using RAMD analysis. J Braz Soc Mech Sci Eng. 2019;*41*:4

[11] Malik S, Tewari PC. Performance modeling and maintenance priorities decision for water flow system of a coal based thermal power plant. Int J Qual Reliabil Manag. 2018;*35*(4):996-1010

[12] Lado A, Singh VV, Ismail KH. Yusuf, I. Performance and cost assessment of repairable complex system with two subsystems connected in series configuration. Int. J. Rel. Appl. 2018;*19*(1):27-42

[13] Chen Y, Meng X, Chen S. Reliability analysis of a cold standby system with imperfect repair and under poisson shocks. Mathematical Problems in Engineering. 2014;**2014**:1-11

[14] Corvaro F, Giacchetta G, Marchetti B, Recanati M. Reliability, availability, maintainability (RAM) study on reciprocating compressors. Petroleum. 2017;*3*(2):266-272

[15] Garg H. An approach for analyzing the reliability of industrial system using fuzzy kolmogrov's differential equations. Arabian Journal for Science and Engineering. 2016a;*40*(3):975-987

[16] Garg H. A novel approach for analyzing the reliability of series-parallel system using credibility theory and different types of intuitionistic fuzzy numbers. Journal of the Brazilian Society of Mechanical Sciences and Engineering. 2016b;*38*(*3*):*1021-1035*

[17] Garg H, Sharma SP. A two phase approach for reliability and maintainability analysis of an industrial system. International Journal of Reliability, Quality and Safety Engineering. 2012;**19**:3

[18] Garg H. Reliability, availability and maintainability analysis of industrial systems using PSO and fuzzy methodology. MAPAN. 2014;**29**(2): 115-129

[19] Kakkar M, Chitkara A, Bhatti J. Reliability analysis of two unit parallel repairable industrial system. Decision Sci. Lett. 2015;**4**(4):525-536

[20] Niwas R, Kadyan MS. Reliability modeling of a maintained system with warranty and degradation. Journal of Reliability and Statistical Studies. 2015;**8** (1):63-75

[21] Negi S, Singh SB. Reliability analysis of non-repairable complex system with weighted subsystems connected in series. Applied Mathematics and Computations. 2015;**262**:79-89

[22] Patil RM, Kothavale BS, Waghmode LY, Joshi SG. Reliability analysis of CNC turning center based on the assessment of trends in maintenance data—a case study. Int J Qual Reliabil Manag. 2017; **34**(9):1616-1638

[23] Tsarouhas PH. Reliability, availability and maintainability (RAM) analysis for wine packaging production line. Int J Qual Reliabil Manag. 2018;**35** (3):821-842. DOI: 10.1108/IJQRM-02-2017-0026

[24] Wang J, Xie N, Yang N. Reliability analysis of a two-dissimilar-unit warm

standby repairable system with priority in use. Commun Stat Theory Methods. 2019. DOI: 10.1080/03610926.2019. 1642488

[25] Wu Q. Reliability analysis of a cold standby system attacked by shocks. Applied Mathematics and Computation. 2012;**218**(23):11654-11673

[26] Wu Q, Wu S. Reliability analysis of two-unit cold standby repairable systems under Poisson shocks. Applied Mathematics and Computation. 2011; **218**(1):171-182

[27] Garg H. An approach for analyzing the reliability of industrial system using fuzzy kolmogrov's differential equations. Arabian Journal for Science and Engineering. 2015;**40**(3):975-987

[28] Kakkar, M., Chitkara, A. and Bhatti, J. (2016). "Reliability analysis of two dissimilar parallel unit repairable system with failure during preventive maintenance", Manage. Sci. Lett., 6(4), pp. 285–296.

[29] Kumar A, Malik SC. Reliability modeling of a computer system with priority to H/W repair over replacement oh H/W and up-gradation of S/W subject to MOT and MRT', *Jordan journal of Mechanical and*. industrial engineering. 2014;**8**(4):233-241

[30] Kumar A, Malik S. Reliability measures of a computer system with priority to PM over the H/W repair activities subject to MOT and MRT. Manage. Sci. Lett. 2015;**5**(1):29-38

[31] Kumar N, Lather JS. Reliability analysis of a robotic system using hybridized technique. J Ind Eng Int. 2018;**14**:443 https://doi.org/10.1007/ s40092-017-0235-5

[32] Kumar A, Pant S, Singh SB. Availability and cost analysis of an engineering system involving subsystems in series configuration. Int J

Qual Reliabil Manag. 2017;**34**(6): 879-894

[33] Suleiman K, Ali UA, Yusuf I. Comparison between four dissimilar solar panel configurations. J Ind Eng Int. 2017;**13**:479. DOI: 10.1007/s40092-017-0196-8

[34] Wang K-H, Kuo C-C. Cost and probabilistic analysis of series systems with mixed standby components. Applied Mathematical Modelling. 2000; **24**:957-967

[35] Wang K-H, Pearn WL. Cost benefit analysis of series systems with warm standby components. Mathematical Methods of Operations Research. 2003; **58**:247-258

[36] Wang K, Hsieh C, Liou C. Cost benefit analysis of series systems with cold standby components and a repairable service station. Journal of quality technology and quantitative management. 2006;**3**(1):77-92

[37] Yen, T,-S and Wang, K.–H. Cost benefit analysis of three systems with imperfect coverage and standby switching failures. Int. Journal of Mathematics in Operational Research. 2018;**12**(2):253-272

Reliability and Risk Analysis

Importance Analysis of Containment Spray System in Pressurized Water Reactor (PWR)

Muhammad Zubair and Priyonta Rahman

Abstract

The basic purpose of the containment spray system (CSS) is to cool the containment atmosphere when the internal pressure of the containment exceeds a certain limit. Water is transferred by a pump from the storage tank via heat exchangers to the overhead spray nozzles in the roof of the containment. This water cools the atmosphere of the containment. In this research, the reliability analysis of CSS has been investigated using fault tree analysis (FTA). The results of the top event probabilities, minimal cut sets (MCS), risk decrease factor (RDF), risk increase factor (RIF), and sensitivity analysis were obtained for the WASH-1400 data base.

Keywords: importance analysis, pressurized water reactor, containment spray system, fault tree analysis, RiskSpectrum

1. Introduction

Nuclear power is a source of sustainable energy, making a significant contribution in the generation of electricity worldwide. Nuclear reactors provide clean energy and is ensured to be safe through the thorough study of nuclear power plants (NPPs) called the probabilistic safety assessment (PSA). PSA is used as an evaluation tool which recognizes the potential risks and accident scenarios resulting in an accident due to the liability of failure of certain components or systems as a whole [1–3]. In an attempt at the aversion of catastrophes, several safety systems are put in place where containment spray system is one among the various redundant safety features in pressurized water reactors (PWRs). The system is aimed at heat removal within the containment when appropriate along with the reduction of radionuclide concentration discharged into the atmosphere. This safety-related system is situated in the auxiliary building and containment of the reactor [4].

A nuclear reactor is provided to take strict safety measures against radioactive contamination of the environment using a diverse arrangement of multiple redundant safety systems [5]. The containment spray system is an engineered safety feature to preserve the integrity of the containment in case of over pressurization. It is designed to bring down the internal peak pressure of the containment by half in a span of 24 hours in case of a loss of coolant accident (LOCA), along with removal of fission products [6].

2. Working of containment spray system

The CSS consists of two identical but independent trains where each train consists of a containment spray pump (CSP), a containment spray heat exchanger (CSHX), a containment spray mini-flow heat exchanger, associated pipes and valves, and containment spray headers located in the upper dome of the containment. A simplified diagram of the CSS is shown in **Figure 1**. The two trains are redundant systems which has the capability of providing 100 percent flow individually during accident conditions and also provides reliability in case one of the trains stop working and makes them testable. In case of maintenance, one of the trains can be shut down and worked on while the other is free to operate.

The CSS in APR-1400 is designed as such that when the internal pressure of the containment surpasses the design limit, a containment spray actuation signal (CSAS) is sent to the pumps to start operation [7]. Once the actuation signal is received, the in-containment refueling water storage tank (IRWST) is used as the suction source for the containment spray system. The CSS pumps discharge water from the IRWST through the set of heat exchangers before sending it to the overhead spray. Once the containment pressure is detected to reach a certain value, the valves open automatically to allow the flow of spray water to the nozzles. The water travels to the spray headers and are divided into small droplets to fall throughout the containment.

Two isolating valves in the system are located in the pipe between the tank and the spray nozzles. The valves are in a closed position under normal operation to isolate the CSS from the rest of the plant. Additional valves are present to ensure the isolation of the overhead spray. The spray nozzles are arranged in concentric circles, at several angles, to ensure adequate coverage of the containment volume resulting in continuous cooling of the reactor system. The droplets from the spray headers cools the atmosphere and the remnants fall into the holdup volume tank (HVT), which is transported back to the IRWST. This certain design of the CSS is maintained in the APR-1400 plant design in countries like Korea, USA, and UAE [8, 9].

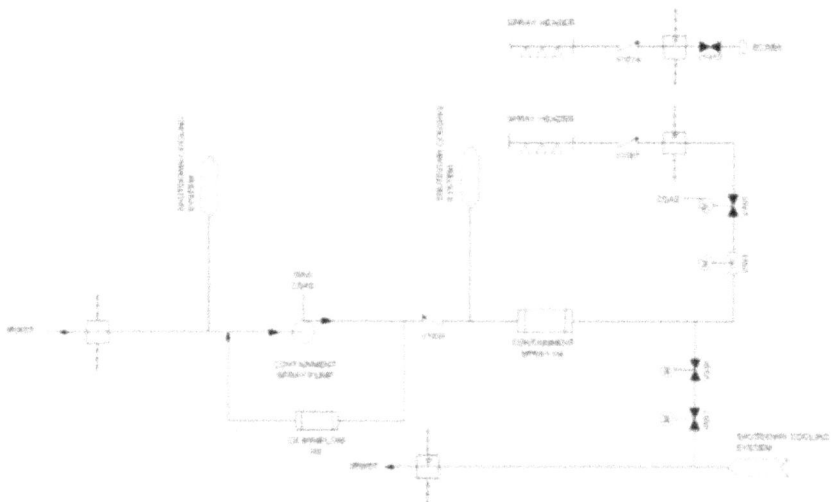

Figure 1.
Containment spray system in APR-1400.

3. Design features of the containment spray system

The containment spray pumps are centrifugal pumps responsible for transporting water from the IRWST to the spray nozzles at the top of the dome. The pumps discharge into the heat exchangers to cool the water. The CSS pumps are identical to the shutdown cooling pumps (SCP) and thus are interchangeable. In case the CSS pumps are inoperative, the SCPs can be used to carry out the operation.

The containment spray heat exchangers are used to remove heat from the containment spray water in the event of an accident. The heat exchangers have a U-tube design with a tube and shell side. The hot water is passed through the tube transferring heat to the cold water in the shell side.

The containment spray headers are located at the top of the containment dome in concentric circles which ensures adequate coverage of the containment building volume and homogeneous distribution of the spray water. The isolation valves in the spray headers control the flow of water into the nozzles and open on the receipt of a CSAS. This reduces the chances of accidental spraying when it is not intended to. The design of the spray nozzles is required to be as such to minimize clogging. They are required to avoid any internal moving parts or restrictions which could possibly interfere with the passage of the flow or restrict it [10].

4. Analysis of CSS fault tree

RiskSpectrum Analysis Tools (RSAT) is a software that enables study of PSA utilizing fault trees. It conducts several analyses such as MCS analysis, uncertainty analysis, importance/sensitivity analysis, and time-dependent analysis. A fault tree was modeled in the RiskSpectrum software according to the fault tree presented in the WASH-1400 report. The report is a 1975 reactor safety study report conducted to assess the accident risks in U.S. commercial nuclear power plants. The unavailability's of the events involved in the fault tree are given in **Table 1** [11].

The fault trees constructed for containment spray system are shown in **Figures 2–4**.

Event		Unavailability (q)
Subsystem A	Subsystem B	
CXVA004X	CXVB004X	1.00E-03
CST100AC	CST100BC	ε
CMV100AC	CMV100BC	1.00E-04
CXVA002X	CXVB002X	1.00E-02
CFLA01AP	CFLB01BP	1.10E-04
CCL1A01G	CCL1B01G	3.00E-04
CM0A01AF	CM0B01BF	ε
CST1A01F	CST1B01F	1.00E-03
CPMA01AA	CPMB01BA	1.00E-03
CPMA01AF	CPMB01BF	1.50E-05
CCVA001C	CCVB001C	1.00E-04
CNZA001P	CNZB001P	1.30E-04

Event		Unavailability (q)
Subsystem A	**Subsystem B**	
GCL01	GCL02	4.60E-03
JD00	JC00	4.10E-05
JK00	JJ00	1.10E-06
CTK0001R		ε
CTK0001L		ε
CVT0001P		4.40E-07
JH00		4.10E-05
JG00		4.10E-05

Note: The symbol "ε" denotes that the unavailability is assumed to be negligible.

Table 1.
Basic event unavailability.

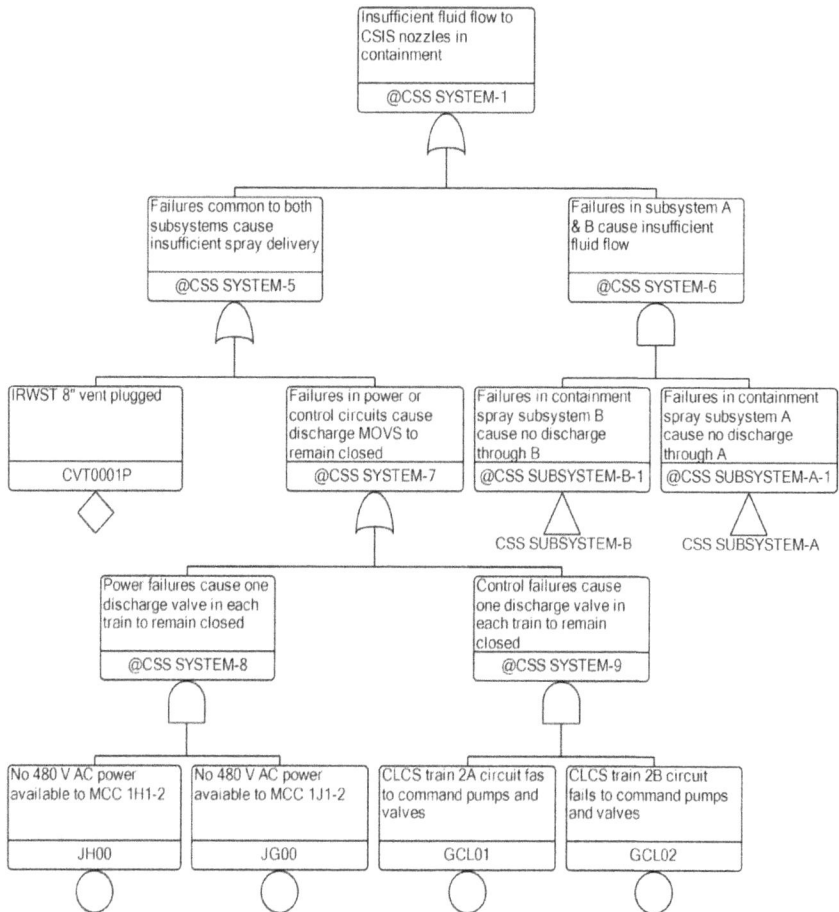

Figure 2.
Containment spray system fault tree (Part 1 of 3).

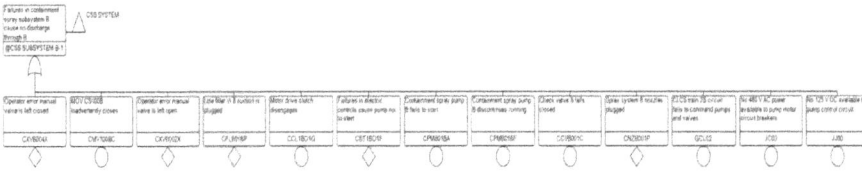

Figure 3.
Containment spray system fault tree (Part 2 of 3).

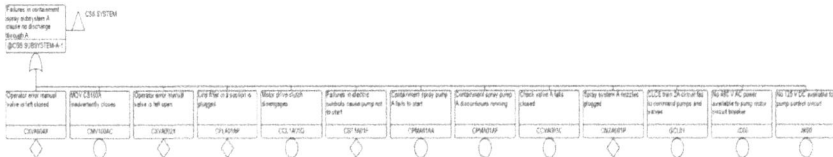

Figure 4.
Containment spray system fault tree (Part 3 of 3).

The MCS analysis conducted on the containment spray system showed that the top event probability (Insufficient fluid flow to CSIS nozzles in containment) is 3.47E-04 with 171 minimal cut sets. The first 10 minimal cut sets against the probability are represented in **Figure 5**. It shows that the minimal cut set with the highest probability of 1.00E-04 is a combination of operator error leaving the manual valve open in both subsystems A and B.

Importance analysis was also conducted on the fault trees which showed the risk decrease factor, risk increase factor, and the sensitivity for each event in **Figures 6–8**, respectively. Risk decrease factor is the reduction in risk if the feature was assumed to be optimized or made perfectly reliable whereas, risk increase factor is the increase in risk if the feature was assumed to be absent or to fail. The graphs below show that manual valves left open in subsystem A due to operator error has the highest RDF, RIF, and sensitivity in the CSS system, followed by the identical event in subsystem B.

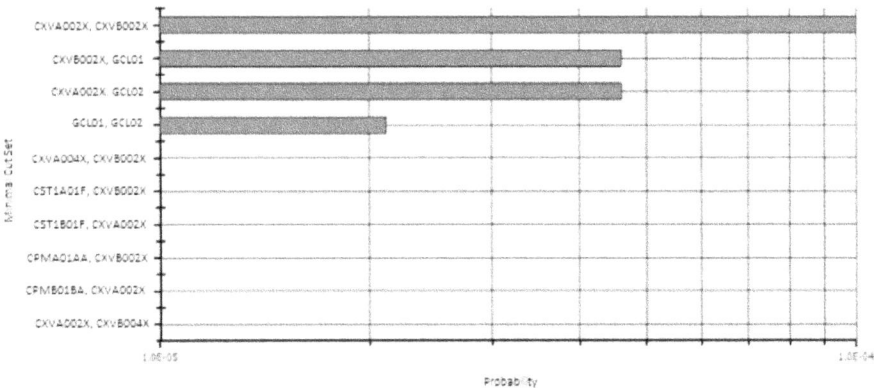

Figure 5.
Top 10 minimal cut sets in a CSS system.

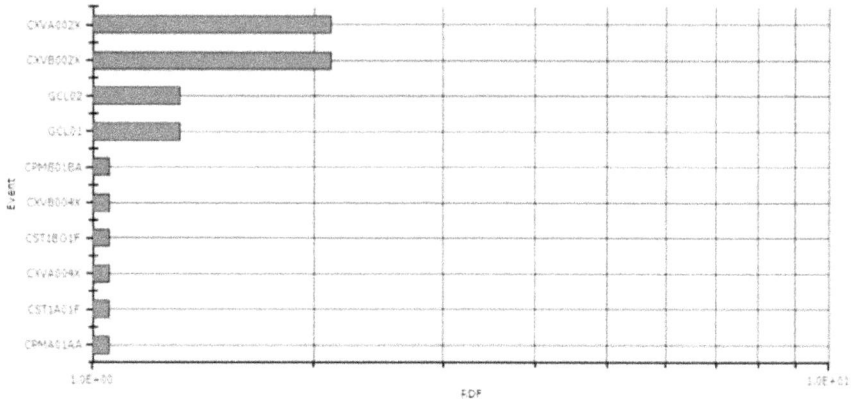

Figure 6.
Basic event RDF.

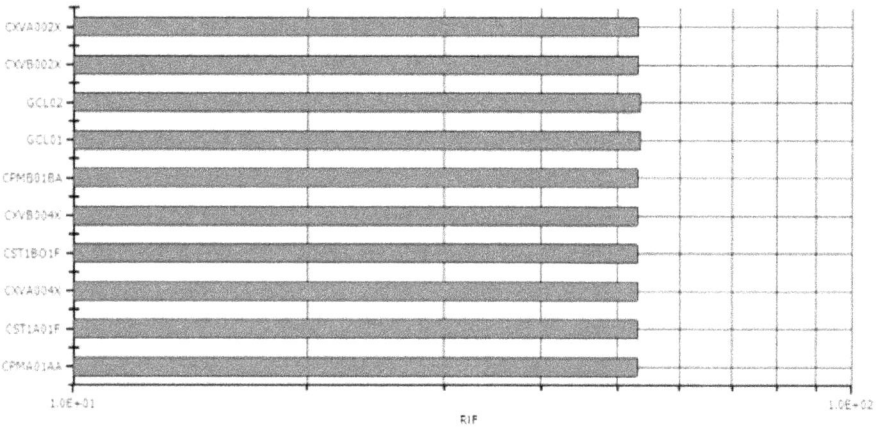

Figure 7.
Basic event RIF.

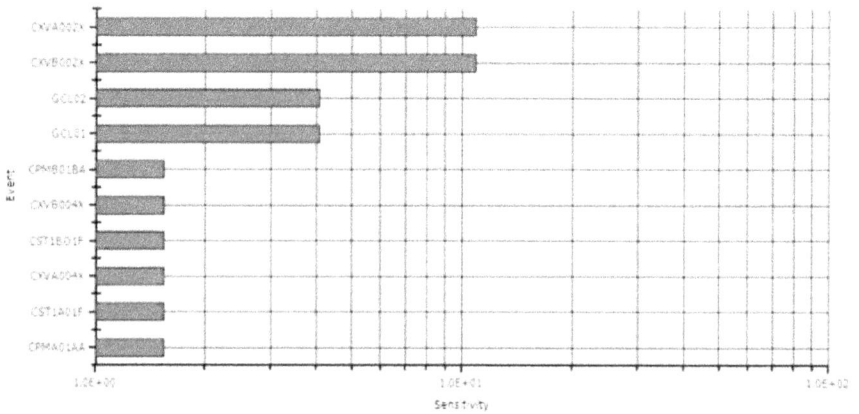

Figure 8.
Basic event sensitivity.

5. Conclusions

Containment spray systems are a crucial part of nuclear power plants as a safety feature. The system is aimed at heat removal within the containment. The CSS consists of two identical, independent trains and is designed as such that when the internal pressure of the containment surpasses the design limit, a containment spray actuation signal starts operation of the CSS. Water from the storage tank is transported to the overhead sprays at the top of the dome from where water is sprayed to ensure adequate coverage of the containment volume resulting in continuous cooling of the reactor system. Analysis of the constructed fault tree in RiskSpectrum from the WASH-1400 report gave in return the top event probability and the list of minimal cut sets. It also showed that the highest RDF, RIF, and sensitivity belonged to the event of operator error where the manual valve is left open in train A.

Acknowledgements

The authors are thankful to university of Sharjah for providing research facilities.

Author details

Muhammad Zubair[1,2]* and Priyonta Rahman[1]

1 Department of Mechanical and Nuclear Engineering, University of Sharjah, Sharjah, UAE

2 Nuclear Energy System Simulation and Safety (NE3S) Research Group, Research Institute of Sciences and Engineering, University of Sharjah, Sharjah, UAE

*Address all correspondence to: mzubair@sharjah.ac.ae

IntechOpen

References

[1] Zubair M, Ababneh A, Ishag A. Station black out concurrent with PORV failure using a generic pressurized water reactor simulator. Annals of Nuclear Energy. 2017;**110**:1081-1090

[2] Zubair M, Ishag A. Sensitivity analysis of APR-1400's reactor protection system by using RiskSpectrum PSA. Nuclear Engineering and Design. 2018;**339**:225-234

[3] Ajit V, Srividya A, Durga K. Reliability and Safety Engineering. London: Springer; 2010. pp. 323-369. DOI: 10.1007/978-1-84996-232-2

[4] US Nuclear Regulatory Commission. APR1400 Design Control Document Tier 1. 2018. Retrieved from https://www.nrc.gov/docs/ML1822/ML18228A647.pdf

[5] Gianni P. Nuclear Safety. 1st ed. Amsterdam: Butterworth-Heinemann; 2006. DOI: https://doi.org/10.1016/B978-0-7506-6723-4.X5000-1

[6] US Nuclear Regulatory Commission. Emergency Core Cooling Systems. 2016. Retrieved from https://www.nrc.gov/docs/ML2005/ML20057E160.pdf

[7] US Nuclear Regulatory Commission. Chapter 11.4 Containment Spray. 2011. Retrieved form https://www.nrc.gov/docs/ML1125/ML11251A006.html

[8] Nuclear Energy Agency. Design description and comparison of design differences between APR1400. Plants. 2018:10 Retrieved from https://ww6w.oecd-nea.org/mdep/documents/TR-APR1400-01%20Design%20Description%20and%20Comparison%20of%20Design%20Differences.pdf

[9] US Nuclear Regulatory Commission. Chapter 6 Engineered Safety Features. 2018. Retrieved from https://www.nrc.gov/docs/ML1822/ML18228A653.pdf

[10] US Nuclear Regulatory Commission. Advanced Power Reactor 1400 (APR1400) Final Safety Evaluation Report. 2018. p. 77-85. Retrieved from https://www.nrc.gov/docs/ML1821/ML18212A092.pdf

[11] U.S. Nuclear Regulatory Commission. Reactor safety study: An assessment of accident risks in U.S. commercial nuclear power plants. In: Appendix. Vol. II. Washington: D.C; 1975

Maintenance and Uncertainty Analysis

Chapter 4

Optimal Maintenance Policy for Second-Hand Equipments under Uncertainty

Ibrahima dit Bouran Sidibe, Imene Djelloul, Abdou Fane and Amadou Ouane

Abstract

This chapter addresses a maintenance optimization problem for re-manufactured equipments that will be reintroduced into the market as second-hand equipments. The main difference of this work and the previous literature on the maintenance optimization of second-hand equipments is the influence of the uncertainties due to the indirect obsolescence concept. The uncertainty is herein about the spare parts availability to perform some maintenance actions on equipment due to technology vanishing. The maintenance policy involves in fact a minimal repair at failure and a preventive repair after some operating period. To deal with this shortcoming, the life cycle of technology or spare parts availability is defined and modeled as a random variable whose lifetimes distribution is well known and Weibull distributed. Accordingly, an optimal maintenance policy is discussed and derived for such equipment in order to overcome the uncertainty on reparation action. Moreover, experiments are then conducted and different life cycle of technologies are evaluated according to their obsolescence processes (accidental or progressive vanishing) on the optimal operating condition.

Keywords: optimal preventive period, indirect obsolescence, minimal repair, second-hand equipment, rejuvenation, virtual age, reliability, residual age

1. Introduction

In a variety of markets and with the rapid economic development, the number of second-hand equipments such as automobiles and high-priced electronic equipment is increasing significantly. Theses equipments tend to degrade with respect to their age and are more likely to fail during their warranty periods than are new equipments. This has generated a stream of parts and goods that can be reconditioned/refreshed to be reused in maintenance actions.

For several decades, many researchers have worked to model and optimize maintenance policies for stochastically degrading production and manufacturing systems. Many interesting and significant results appeared in the literature. The initial framework for preventive maintenance (PM) is due to Barlow and hunter in their seminal paper [1]. Subsequently, a large variety of mathematical models appeared in the literature for optimal maintenance policies design and implementation. For a review on the topic, the reader is referred to [2–5], and the references therein. Recently Nakagawa and his coauthors proposed new models dealing with finite time horizon [6, 7]. Lugtighei and his co-authors also achieved a review of maintenance models [8].

The recent expansion of transaction volume on second-hand market has therefore made grown the potential benefit and research interest for businesses and equipments on such market through a better modeling of additional services such as warranty and maintenance optimization. Accordingly, several researches are performed in order to fit adaptive warranty policy for second-hand equipments. In fact, an analysis of warranty cost was discussed in [9] while Shafiee and al. proposed an approach to determinate an optimal upgrade level for second-hand equipment according to the overhaul cost structure in [10]. In the same way, seminal research was also conducted on optimizing maintenance or replacement policy for second-hand equipment in recent decade. Therefore, optimal maintenance policies were adjusted for second-hand equipments in [11, 12]. In which the authors established maintenance policies which ensure an optimal preventive repair period for second-hand equipment based on the cost and hypotheses on its initial age (deterministic or random). However, the maintenance models proposed in above references assume that the technology and spare parts to repair equipment remains available during optimization period.

The recent economic downturn combined with the rapid development of technology drawn a new business ecosystem with a new relationship between the producers and the consumers. The new ecosystem requires handy equipments with more innovation and cheaper. This situation affects deeply the industrial design and involves additional research cost and reduce the cycle life of product in manufacturer industry. To deal with this cost, the producers have suited their policy by adjusting the sale volume through repeat purchase in order to keep their profit margin. The repeat purchase involves making product with deliberated short life or useful life. Short products life cycle implies premature obsolescence. The premature obsolescence can be ones or combination of planned obsolescence, indirect obsolescence, incompatibility obsolescence, Style obsolescence. In planned obsolescence, the equipment fails systematically after some durations and without possibility to repair. Moreover the indirect obsolescence involves a deliberate unavailability of spare parts. Without spare parts, the repair execution becomes impossible to do in practice. We note that the consequences of premature obsolescence reduce the chance to perform maintenance with the time in practice. This situation makes equipment unrepairable and the execution of older maintenance policy unlikely. Accordingly, this situation arouses issues about our manner to think the maintenance optimization approach regardless the type of equipment (new or second-hand). This chapter is therefore going to highlight the drawback and impact of indirect obsolescence on a maintenance policy optimizing through the random modeling of maintenance execution in practice.

Based on the previous notes that a stochastic maintenance strategy is herein proposed and discussed under indirect obsolescence. The maintenance cost is modeled and analyzed according to the residual age of technology and its vanishing process (accidental, progressive).

The remainder of chapter is organized as follows. Next section explains clearly the problematic of maintenance policy. This section defines the repair strategy and introduces the nature of each repair. In Section 3, a mathematical model of maintenance cost is proposed with some explanations on the cost. In section 4, we discuss the optimality condition with respect to each repair cost and cost parameters. We finish this section by numerical experiment in which the maintenance policy is analyzed through the technology vanishing process.

2. Problem description

We consider a smart maintenance policy for second-hand equipments under an uncertainty on the execution of repair due to the unavailability of required spare parts or technologies. The equipment is bought on second-hand market and rejuvenated for a safe operation. The rejuvenation has a cost and allows to reach a required reliability for second-hand equipment.

The considered equipments are assumed to operates under some uncertainties. To perform their tasks correctly, the equipments are going to be repaired minimally at failure and preventively after some operating period. Each repair involves a cost which depends on the type of repair.

The minimal repair takes place at failure. This reparation does not impact equipment age and allows to maintain equipment failure risk at the level before it fails. Meanwhile, the preventive repair equipment is performed after some operating period. This repair consists of a soft overhaul of equipment and allows to reduce the equipment age and then the failure risk. The preventive repair requires higher cost and uses more spare parts than minimal repair. All repairs need new spare parts to replace or repair faulty components anywhere and anytime on equipment in order that the equipment gets a minimal required failure risk to operate in safe condition. The replacement of faulty or failed components involves new one available. However, this availability becomes uncertain with the indirect obsolescence.

This uncertainty is therefore considered in our works. Accordingly, the useful life of the technology or the availability period of spare parts behind equipments is considered as a random value with a continuous probability distribution. To derive a cost of maintenance policy for such equipments under uncertainty of technology, the chance to perform reparation is evaluated and integrated in the models. The appraisal of the probability to perform repair depends on the residual life distribution of technology and its probability distribution function.

The next section proposes a mathematical model for the maintenance policy optimization under uncertainty on the technology availability to ensure the repair.

3. Mathematical model of maintenance cost

The second-hand equipment with certain ages u is considered in this chapter. In fact an equipment is bought on second-hand market with an initial age at price $C_{ac}(u)$. The cost $C_{ac}(u)$ stands for the acquisition price whose value depends on the age at acquisition and the technology. if we assume that the new equipment costs C_{new} then the acquisition price function C_{ac} has to respect some mathematical properties such as

$$C_{ac}(0) = C_{new}, \tag{1}$$

$$\frac{d}{du}C_{ac}(u) \leq 0. \tag{2}$$

$$\lim_{u \to \infty} C_{ac}(u) = 0, \tag{3}$$

Roughly speaking, an equipment at age $u = 0$ is bought at the price of new Eq. (1). Moreover, the acquisition price C_{ac} remains non-increasing function with respect to acquisition age (u) (Eq. (2)) and becomes null for older equipment Eq. (3).

In addition, the bought equipment is overhauled before operating. This overhaul consists of deep repair for the equipment in order to get a required threshold in reliability by rejuvenation. The rejuvenation involves a safe operating condition costs C_{rej}. This cost C_{rej} is function of initial age u and the desired rejuvenation on the equipment age u_f. $C_{rej}(u, u_f)$ stands for the rejuvenation cost and respect also some mathematical properties defined as follows:

$$\begin{cases} C_{rej}(u, u_f = u) = 0, \\ C_{rej}(u, u_f = 0) = \infty, \end{cases} \tag{4}$$

Accordingly, the rejuvenation allow to reduce the age of equipment from u to u_f such as $u \geq u_f$. To fit reality, we assume that having new equipment by rejuvenation is impossible and non-rejuvenation costs anything in practice Eq. (4). In addition the rejuvenation cost increases in function of initial age u Eq. (5) while it decreases with the final age u_f Eq. (6)

$$\frac{\partial}{\partial u} C_{rej}(u, u_f) > 0, \tag{5}$$

$$\frac{\partial}{\partial u_f} C_{rej}(u, u_f) < 0. \tag{6}$$

3.1 Preventive repair cost

The preventive repair involves more spare parts and duration to diagnose and to replace failed components. This repair is performed after each operating period with T_k length. Each preventive repair is assumed imperfect but better than the minimal repair. Herein, the preventive repair is expressed by age reduction model with infinite memory [13]. The preventive action allows to reduce the virtual age of equipment by a multiplicative factor α_k. Each preventive repair involves a cost $C_p(k, \alpha_k)$ and also spends a duration which is negligible relatively to the length of operation period.

Moreover the preventive cost depends on the age reduction according to the factor α_k. The preventive repair cost is assumed non-decreasing function with respect k and denotes by $C_p(k)$.

$$C_p(k, \alpha_k) = C_p(k). \tag{7}$$

The virtual age reduction due to preventive repair is performed according to α_k multiplicative factor. Therefore in case of multiplicative factor the reduction after k^{th} preventive repair is $\alpha_k Age_{k-1}$. Therefore, the virtual age of equipment at the end of kth period

$$Age_k = \alpha_k Age_{k-1} + T_k, \tag{8}$$

where Age_{k-1} represents the virtual age of equipment before the kth preventive repair. Therefore k^{th} preventive impact the age of equipment by factor α_k. Accordingly, the age of equipment after the kth preventive becomes Age_k.

Proposition 1. *The virtual age of equipment at beginning of the kth operating period is therefore equal to:*

$$\begin{cases} Age_0 = \alpha_0 u, & \text{if } k = 1 \\ Age_{k-1} = \left(\prod_{j=0}^{k-1} \alpha_j \right) u + \sum_{m=1}^{k-1} \left(\prod_{j \geq m}^{k-1} \alpha_j \right) T_m, & \text{otherwise} \end{cases} \tag{9}$$

with $\alpha_{rej} = \alpha_0$.

Proof: From Eq. (8) we deduce each virtual age as follows

$$Age_{k-1} = \alpha_{k-1} Age_{k-2} + T_{k-1} \tag{10}$$

for different values of k suxh as k in $\{k-2, k-3, \ldots, 1\}$, we obtain then

$$Age_{k-2} = \alpha_{k-2} Age_{k-3} + T_{k-2} \tag{11}$$

$$\cdots = \cdots\cdots\cdots$$

$$Age_2 = \alpha_2 Age_1 + T_2 \tag{12}$$

$$Age_1 = \alpha_1 Age_0 + T_1 \tag{13}$$

$$Age_0 = \alpha_0 u \tag{14}$$

The result of proposition (1) is deduced by recursive replacement of each Age_i from bottom to top. Therefore we obtain

$$
Age_{k-1} = (\alpha_{k-1}\alpha_{k-2}\cdots\alpha_1\alpha_0)u
$$
$$
+ (\alpha_{k-1}\alpha_{k-2}\cdots\alpha_1)T_1 + (\alpha_{k-1}\alpha_{k-2}\cdots\alpha_2)T_2
$$
$$
+ \cdots\cdots\cdots \tag{15}
$$
$$
+ (\alpha_{k-1})T_{k-2} + T_{k-1}
$$

this can be rewritten as follows

$$Age_{k-1} = \left(\prod_{j=0}^{k-1} \alpha_j \right) u + \sum_{m=1}^{k-1} \left(\prod_{j \geq m}^{k-1} \alpha_j \right) T_m, \quad \text{with} \quad k > 1 \tag{16}$$

3.2 Minimal repair cost

The minimal repair takes place at failure. The equipment after such repairs is considered As Bad As Old (ABAO). The cost of such repair during the kth operating period with T_k length is denoted by $C_m(k)$ and depends on the expected number of failures and the unit cost per repair. The minimal repair cost on k^{th} operating period is

$$C_m(k) = C_{m|k} \left(\int_{Age_{k-1}}^{Age_{k-1}+T_k} \lambda(t)dt \right), \tag{17}$$

where $C_{m|k}$ and $\int_{Age_{k-1}}^{Age_{k-1}+T_k} \lambda(t)dt$ stand respectively for the unit minimal repair cost and the expected number of failures. Moreover $C_{m|k}$ and $\lambda(t)$ are non-decreasing function with respect respectively to k and t. From Eq.(17), we deduce the mathematical formula of cost $C_m(k)$ by integration

$$C_m(k) = C_{m|k} \left(\Lambda(Age_{k-1} + T_k) - \Lambda(Age_{k-1}) \right), \tag{18}$$

with $\Lambda(t)$ the cumulative failure risk function of equipment. Eq.(18) depends on the age reduction Age_{k-1} process according to the effect of preventive repair.

3.3 Repair cost during period k

The repair cost during k^{th} period involves the both repairs (minimal and preventive) from Eq. (7) and Eq. (18). The repair cost due to operating on period k is derived as follows:

$$C(k) = C_m(k) + C_p(k-1). \tag{19}$$

Eq.(19) holds under hypothesis that the technologies or spare parts are always available during the kth. The required technology to repair our equipment is available with uncertainty. This uncertainty is governed by the residual life cycle of the equipment technology and its probability distribution. In fact to perform reparation during period k^{th}, the technology has to be available. This availability is not certain and requires a probability due to uncertainty on the spare parts. Therefore for a given technology characterized by its life cycle Y, we define the probability to perform these repairs by p_k and compute it by use of the residual life cycle distribution as follows

$$p_k = Prob\left(Y \geq \tilde{T}_{k-1} + Y_{ac} | Y \geq Y_{ac}\right), \tag{20}$$

where Y_{ac} stands for the age of technology at acquisition date of equipment and $\tilde{T}_{k-1}$ represents the cumulative operating duration. Therefore

$$\tilde{T}_{k-1} = \sum_{j=1}^{k-1} T_j, \tag{21}$$

and

$$p_k = \frac{Prob\left(Y \geq \tilde{T}_{k-1} + Y_{ac}\right)}{Prob(Y \geq Y_{ac})}. \tag{22}$$

Based on Eq. (22), we derive the repair cost which takes into account of the uncertainty of technology and the probability to perform each repair during period k by

$$\tilde{C}(k) = \frac{C(k)}{p_k}. \tag{23}$$

The cost $\tilde{C}(k)$ remains more realistic to evaluate the expected repair cost. This later cost integrates the cost of repairs and the probability to perform its in practice under some uncertainties. Through Eq.(23), we ensure in addition that when the technology is certainly available both cost are the same. However for unavailable technology during period k the cost $\tilde{C}(k)$ tends toward infinity in order to highlight the unrepairable case of equipment due to unavailability of spare parts.

3.4 Total maintenance cost

The total cost due to maintenance policy is function of acquisition price $C_{ac}(u)$ of equipment. In addition, the total cost implies also all repairs costs regardless its nature (preventive or minimal). Indeed, the total maintenance cost represents the sum of all costs (acquisition, rejuvenation and reparation). For a given number of operating sequences n, the total maintenance cost of our maintenance policy is derived and written as follows:

$$C_T(n) = \sum_{k=1}^{n} \tilde{C}(k) + C_{ac}(u). \tag{24}$$

Based on Eq. (25), we deduce the total maintenance cost per unit time as the ratio between $C_T(n)$ and the length or duration of operating period $\sum_{k=1}^{n} T_k$.

$$\tilde{C}_T(n) = \frac{C_T(n)}{\sum_{k=1}^{n} T_k}.$$

(25)

Remaining of the chapter is going to make a discussion on the conditions which ensure the optimality of the total maintenance cost or the total maintenance cost per unit time defined. This analysis is going to be done through some propositions.

4. Analysis of maintenance cost

The optimality of the maintenance is going to be deeply discussed throughout some parameters. In fact the cost model proposed in Eq. (24) and Eq. (25) depend on several parameters from equipment design, the efficiency of repair action and environment through the technology useful life. This section expects to derive optimal conditions.

Proposition 2. *The optimal age u to acquire a second-hand equipment according to our maintenance policy verifies*

$$-\frac{\partial C_{ac}(u)}{\partial u} = \sum_{k=1}^{n} \frac{1}{p_k} \left(C_{m|k} \left(\lambda(Age_{k-1} + T_k) - \lambda(Age_{k-1}) \right) \prod_{j=0}^{k-1} \alpha_j + \frac{\partial C_p(k-1)}{\partial u} \right)$$

(26)

Proof: The cost model is derived with respect to u. We note that the derivative function of the total cost per unit time with respect to u is equivalent to $\frac{\partial}{\partial u} C_T(n)$. Then

$$\frac{\partial}{\partial u} C_T(n) = \frac{\partial}{\partial u} \left(\sum_{k=1}^{n} \tilde{C}(k) + C_{ac}(u) \right),$$

$$= \left(\sum_{k=1}^{n} \frac{\partial \tilde{C}(k)}{\partial u} + \frac{\partial C_{ac}(u)}{\partial u} \right),$$

(27)

from Eq. (23), we deduce $\frac{\partial \tilde{C}(k)}{\partial u} = \frac{1}{p_k} \frac{\partial C(k)}{\partial u}$. Then

$$\frac{\partial C(k)}{\partial u} = \frac{\partial}{\partial u} \left(C_m(k) + C_p(k-1) \right),$$

$$= \frac{\partial C_m(k)}{\partial u} + \frac{\partial C_p(k-1)}{\partial u},$$

with

$$\frac{\partial C_m(k)}{\partial u} = \left(C_{m|k} \frac{\partial Age_{k-1}}{\partial u} \right) \left(\lambda(Age_{k-1} + T_k) - \lambda(Age_{k-1}) \right)$$

(28)

Accordingly, the derivation of cost due to the kth operating cycle is written as follows

$$\frac{\partial C(k)}{\partial u} = C_{m|k} \left(\frac{\partial Age_{k-1}}{\partial u} \right) \left(\lambda(Age_{k-1} + T_k) - \lambda(Age_{k-1}) \right) + \frac{\partial C_p(k-1)}{\partial u},$$

(29)

and with uncertainty, we obtain then

$$\frac{\partial \tilde{C}(k)}{\partial u} = \frac{1}{p_k}\left(C_{m|k}\frac{\partial Age_{k-1}}{\partial u}\left(\lambda(Age_{k-1}+T_k)-\lambda(Age_{k-1})\right)+\frac{\partial C_p(k-1)}{\partial u}\right) \quad (30)$$

with $\frac{\partial Age_{k-1}}{\partial u} = \prod_{j=0}^{k-1}\alpha_j$ according to Eq.(9). The derivative function of total cost $\frac{\partial C_T(n)}{\partial u} = 0$ or total cost per unit time $\frac{\partial \tilde{C}_T(n)}{\partial u} = 0$ are deduced as follows:

$$\frac{\partial C_T(n)}{\partial u} = 0 \Leftrightarrow \frac{\partial \tilde{C}_T(n)}{\partial u} = 0$$

and

$$\frac{\partial C_T(n)}{\partial u} = \sum_{k=1}^{n}\frac{1}{p_k}\left(C_{m|k}\left(\lambda(Age_{k-1}+T_k)-\lambda(Age_{k-1})\right)\prod_{j=0}^{k-1}\alpha_j+\frac{\partial C_p(k-1)}{\partial u}\right) \\ +\frac{\partial C_{ac}(u)}{\partial u} \quad (31)$$

therefore posing the right-hand term of Eq.(31) equals to zero allows to get Eq.(26). Moreover the preventive repair cost is often independent of the acquisition age u of second-hand equipment. In this case, the optimal condition on acquisition age u is

$$\frac{\partial C_{ac}(u)}{\partial u} = -\sum_{k=1}^{n}\frac{1}{p_k}\left(C_{m|k}\left(\lambda(Age_{k-1}+T_k)-\lambda(Age_{k-1})\right)\prod_{j=0}^{k-1}\alpha_j\right). \quad (32)$$

Another way, the proposed maintenance policy involves a sequence of preventive repairs. These repairs are performed after some operating periods $T_{i,i=1,\cdots,n}$ whose lengths ensure a minimal maintenance cost per unit time. The next proposition establishes then the conditions on each operation period $T_{i,i=1,\cdots,n}$ before preventive repair for an optimal use of equipment under our maintenance policy.

Proposition 3. *The optimal operating duration before the ith preventive repair is fulfilled for a T_i such as*

$$\sum_{k=1}^{n}\frac{\partial}{\partial T_i}\left(\frac{C(k)}{p_k}\right) = \tilde{C}_T(n) \Rightarrow \sum_{k=1}^{n}\left(\frac{1}{p_k}\frac{\partial C_m(k)}{\partial T_i}-\frac{C(k)}{p_k^2}\frac{\partial p_k}{\partial T_i}\right) = \tilde{C}_T(n) \quad (33)$$

with

$$\frac{\partial C_m(k)}{\partial T_i} = \begin{cases} 0, & \text{if } i>k, \\ C_{m|k}\lambda(Age_{k-1}+T_k), & \text{if } i=k, \\ C_{m|k}\left(\lambda(Age_{k-1}+T_k)-\lambda(Age_{k-1})\right)\frac{\partial Age_{k-1}}{\partial T_i}, & \text{otherwise}, \end{cases} \quad (34)$$

where $\frac{\partial Age_{k-1}}{\partial T_i} = \prod_{j\geq i}^{k-1}\alpha_j$ and

$$\frac{\partial p_k}{\partial T_i} = \begin{cases} 0, & \text{if } i>k \\ -\dfrac{f_{tech}\left(\tilde{T}_{k-1}+Y_{ac}\right)}{\overline{F}_{tech}(Y_{ac})}, & \text{otherwise}. \end{cases} \quad (35)$$

where f_{tech} and $\overline{F}_{tech}$ stand for density and reliability functions of spare parts or technology availability during optimizing period.

Proof: Each optimal T_i verifies next equation

$$\frac{\partial \tilde{C}_T(n)}{\partial T_i} = 0, \tag{36}$$

this is equivalent to

$$\frac{\partial \tilde{C}_T(n)}{\partial T_i} = \frac{\partial}{\partial T_i}\left(\frac{C_T(n)}{\tilde{T}_n}\right), \tag{37}$$

then Eq.(36) becomes

$$\frac{\partial C_T(n)}{\partial T_i} = \tilde{C}_T(n), \tag{38}$$

therefore the left-hand term of Eq.(38) is equal to

$$\frac{\partial C_T(n)}{\partial T_i} = \frac{\partial}{\partial T_i}\left(\sum_{k=1}^{n}\frac{C(k)}{p_k} + C_{ac}(u)\right),$$

$$= \left(\frac{\frac{\partial C(k)}{\partial T_i}p_k - C(k)\frac{\partial p_k}{\partial T_i}}{p_k^2} + \frac{\partial C_{ac}(u)}{\partial T_i}\right),$$

$$= \sum_{k=1}^{n}\left(\frac{1}{p_k}\frac{\partial C(k)}{\partial T_i} - \frac{C(k)}{p_k^2}\frac{\partial p_k}{\partial T_i}\right)$$

the derivative function of $\frac{\partial \tilde{C}_T}{\partial T_i}$ depends on the $\frac{\partial C(k)}{\partial T_i}$ and $\frac{\partial p_k}{\partial T_i}$. In fact

$$\frac{\partial C(k)}{\partial T_i} = \frac{\partial C_m(k)}{\partial T_i} + \frac{\partial C_p(k-1)}{\partial T_i}$$

and

$$\frac{\partial C_p(k-1)}{\partial T_i} = 0$$

then Eq.(33) is deduced based on

$$\frac{\partial C(k)}{\partial T_i} = \frac{\partial C_m(k)}{\partial T_i}$$

with

$$\frac{\partial C_m(k)}{\partial T_i} = \begin{cases} 0, & \text{if } i > k, \\ C_{m|k}\lambda\left(Age_{k-1} + T_k\right), & \text{if } i = k, \\ C_{m|k}\left(\lambda\left(Age_{k-1} + T_k\right) - \lambda\left(Age_{k-1}\right)\right)\frac{\partial Age_{k-1}}{\partial T_i}, & \text{otherwise,} \end{cases} \tag{39}$$

where $\frac{\partial Age_{k-1}}{\partial T_i} = \prod_{j \geq i}^{k-1}\alpha_j$ is deduced from Eq.(9) of Proposition (1). In addition, the derivative $\frac{\partial p_k}{\partial T_i}$ is

$$\frac{\partial p_k}{\partial T_i} = \frac{\partial}{\partial T_i} Prob\left(Y > \tilde{T}_{k-1} + Y_{ac} | Y > Y_{ac}\right),$$

$$= \frac{\partial}{\partial T_i} \frac{1 - F_{tech}\left(\tilde{T}_{k-1} + Y_{ac}\right)}{1 - F_{tech}(Y_{ac})}, \tag{40}$$

the derivative of p_k with respect T_i is

$$\frac{\partial p_k}{\partial T_i} = \begin{cases} 0 & \text{si } i \geq k, \\ -\dfrac{f_{tech}\left(\tilde{T}_{k-1} + Y_{ac}\right)}{\overline{F}_{tech}(Y_{ac})} & \text{otherwise.} \end{cases} \tag{41}$$

Therefore the derivative of total cost per unit time with respect is deduced from Eq.(39) and Eq.(40) by

$$\frac{\partial}{\partial T_i}\left(\frac{C(k)}{p_k}\right) = \begin{cases} 0, & \text{if } i > k, \\[2mm] \dfrac{C_{m|k}}{p_k} \lambda\left(Age_{k-1} + T_k\right), & \text{if } i = k, \\[3mm] \left(\dfrac{C_{m|k}}{p_k}\left(\lambda\left(Age_{k-1} + T_k\right) - \lambda\left(Age_{k-1}\right)\right)\right) \displaystyle\prod_{j \geq i}^{k-1} \alpha_j \\[3mm] + \dfrac{f_{tech}\left(\tilde{T}_{k-1} + Y_{ac}\right)}{1 - \overline{F}_{tech}(Y_{ac})}\left(C_m(k) + C_p(k-1)\right)\dfrac{1}{p_k^2}, & \text{otherwise.} \end{cases} \tag{42}$$

The next proposition derives the optimal condition for α_i which stands for the age reduction factor due to the preventive repair if it takes place.

Proposition 4. *Each optimal reduction factor* $\alpha_{i,i=1,\cdots,n-1}$ *is derived as a solution of next equation*

$$\sum_{k=1}^{n} \frac{\partial C_m(k)}{\partial \alpha_i} = -\frac{\partial C_p(i)}{\partial \alpha_i} \tag{43}$$

Proof: The proof of this proposition results of the calculation of the derivative function with respect to α_i and posing equal to zero. This implies

$$\frac{\partial}{\partial \alpha_i}\left(\frac{\tilde{C}_T(n)}{\tilde{T}_n}\right) = 0, \quad \Rightarrow$$

$$\frac{\partial C_T(n)}{\partial \alpha_i} = 0,$$

$$\frac{\partial C_T(n)}{\partial \alpha_i} = \frac{\partial}{\partial \alpha_i} \sum_{k=1}^{n}\left(C_m(k) + C_p(k-1)\right)$$

$$= \sum_{k=1}^{n} \frac{\partial C_m(k)}{\partial \alpha_i} + \frac{\partial C_p(k-1)}{\partial \alpha_i}$$

$$= \sum_{k=1}^{n}\left(\frac{\partial C_m(k)}{\partial \alpha_i}\right) + \frac{\partial C_p(i)}{\partial \alpha_i}$$

therefore

$$\frac{\partial C_T(n)}{\partial \alpha_i} = 0, \quad \Rightarrow$$

$$\sum_{k=1}^{n} \left(\frac{\partial C_m(k)}{\partial \alpha_i} \right) = -\frac{\partial C_p(i)}{\partial \alpha_i}$$

with

$$\frac{\partial C_m(k)}{\partial \alpha_i} = C_{m|k} \left(\lambda (Age_{k-1} + T_k) - \lambda (Age_{k-1}) \right) \frac{\partial Age_{k-1}}{\partial \alpha_i} \quad (44)$$

and

$$\frac{\partial Age_{k-1}}{\partial \alpha_i} = \begin{cases} 0, & i > k-1, \\ \left(\prod_{j=0, j\neq i}^{k-1} \alpha_j \right) u + \sum_{m=1}^{k-1} \prod_{j\geq m, j\neq i}^{k-1} \alpha_j T_m, & i \leq k-1. \end{cases} \quad (45)$$

The last proposition deals with the optimality according to the number n of operating periods. We note that n is integer and the derivative with respect to n is not possible. However to make comparison, the maintenance cost per unit time is going to be evaluate for $n-1$, n, and $n+1$ number of operating periods.

Proposition 5. *A number of maintenance periods n is optimal according to our maintenance policy if it verifies*

$$\begin{cases} U_n > \dfrac{\tilde{T}_{n+1}}{\tilde{T}_n}, \\ U_{n-1} < \dfrac{\tilde{T}_n}{\tilde{T}_{n-1}}. \end{cases} \quad (46)$$

where U_n is sequence such as $U_n = \frac{\check{C}_T(n+1)}{\check{C}_T(n)}$ ..

Proof: n is integer and at the optimal the total maintenance cost ensure the next inequalities

$$\begin{cases} \dfrac{\check{C}_T(n+1)}{\tilde{T}_{n+1}} > \dfrac{\check{C}_T(n)}{\tilde{T}_n} & (a) \\ \dfrac{\check{C}_T(n-1)}{\tilde{T}_{n-1}} > \dfrac{\check{C}_T(n)}{\tilde{T}_n} & (b) \end{cases} \quad (47)$$

we derive then the optimal condition for n as follows

$$\begin{cases} \dfrac{\check{C}(n+1)}{\check{C}(n)} > \dfrac{\tilde{T}_{n+1}}{\tilde{T}_n} & (a') \\ \dfrac{\check{C}(n)}{\check{C}(n-1)} < \dfrac{\tilde{T}_n}{\tilde{T}_{n-1}} & (b') \end{cases} \quad (48)$$

Eq.(48) depicts that when U_n is increasing sequence such $U_0 < 1$ then the optimal n is unique.

$$\begin{cases} \dfrac{\tilde{T}_{n+1}}{\tilde{T}_n} = 1 + \dfrac{T_{n+1}}{\tilde{T}_n} \approx 1, \\[3mm] \dfrac{\tilde{T}_n}{\tilde{T}_{n-1}} = 1 + \dfrac{T_n}{\tilde{T}_{n-1}} \approx 1. \end{cases} \qquad (49)$$

In case of Eq.(49), the optimal n verifies both conditions $U_n > 1$ and $U_{n-1} < 1$. However, the complexity of the cost function structure makes the monotonicity analysis of U_n difficult in practice. Therefore the optimization in numerical experiment is going to perform for given number n of operating periods.

Next section propose a numerical experiment in order to analyze the optimal solution under uncertainty on the technology or spare parts availability.

4.1 Numerical experiments

To make experiment, a second-hand equipment is considered with Weibull distribution as it lifetime distribution. The parameter of the Weibull distribution function are $\beta = 2.1$ and $\eta = 100$ which respectively stand for the shape and scale parameters. Additional, the second-hand equipment is assumed from technology whose vanishing may be accidental or progressive. In case of accidental vanishing, the technology life cycle is distributed according to a exponential distribution function while the technology life cycle follows Weibull distribution function with a shape parameter $\beta_{tech} > 1$ in case of progressive [14]. **Table 1** depicts the all parameters for a numerical experiment such as the acquisition price of second-hand equipment and the unit costs to perform repair.

To compare technology effect on our maintenance policy, two hypotheses are made. First, equipments are made from newer technology at beginning of its life cycle. Second, we assume that the equipment technology is at the end of its life cycle. In fact the life cycle of technology distribution are also modeled by Weibull probability distribution with $\eta_{tech} = 1000$ as the scale parameter and the shape parameters equal to $\beta_{tech} = 1$ and $\beta_{tech} = 4.1$ respectively for accidental and progressive vanishing. We note solving the optimization issues remains complex due to the number and the nature of policy parameters.

To deal with the complexity of problem, we focus on optimization and analysis of the first operating period. The results of this analysis are displayed in the next figures. **Figure 1** depicts the evolution of optimal rejuvenation ratio α_0 applied on equipment before operating. This figure highlights a period of acquisition age in which the equipment does not need rejuvenation. The length of this period depends on the age of technology and the vanishing process. Therefore, the length of this period is longer for equipments based on older technology with progressive

Equipment lifetime	Weibull distribution	$W(\beta = 100, \eta = 2.1)$
Technology life cycle	Accidental vanishing	$W(\beta = 1, \eta = 1000)$
Technology life cycle	Progressive vanishing	$W(\beta = 4.01, \eta = 1000)$
Acquisition price of new	$C_{new} = 100000$	
Acquisition price function	$C_{ac}(u) = \frac{C_{new}}{1+u}$	
Cm_0	500	
Cp_0	1000	

Table 1.
Numerical Experiments parameters.

vanishing while its remains short for equipments based on newer technology. Moreover when vanishing process is accidental, the length of this period reach 15.79 unit of time. From **Figure 1**, we deduce that the optimal α_0 decreases with the initial acquisition age. For progressive vanishing technology, the required rejuvenation to optimally operate remains more important when the life cycle of technology is at beginning than ending. This results from the high probability of technology unavailability. Therefore, when the spare parts availability is uncertain the maintenance policy recommends a soft rejuvenation while a deep rejuvenation is needed otherwise. Another way, for the same age of technology at acquisition the rejuvenation in accidental vanishing less important than in progressive vanishing according to **Figure 1**.

Figure 2 presents the optimal operating period length versus initial acquisition age. **Figure 2** underlines that the optimal operating period decreases also with the acquisition age regardless the life cycle distribution of technology. Therefore the maintenance policy recommends short operating length with ending technology. In addition, the maintenance policy requires early preventive repair in case of progressive vanishing than accidental. This allows to reduce the uncertainty effect on the chance to perform repairs.

Figure 3 presents and allows to derive the optimal acquisition age to get the minimal operating cost per unit time during the first period of operation. The optimal cost per unit time reach minimal for accidental, progressive with $Y_{ac} = 100$ and $Y_{ac} = 1000$ at respectively to $45.71, 42.89,$ and $55, 63$. This involves that the use

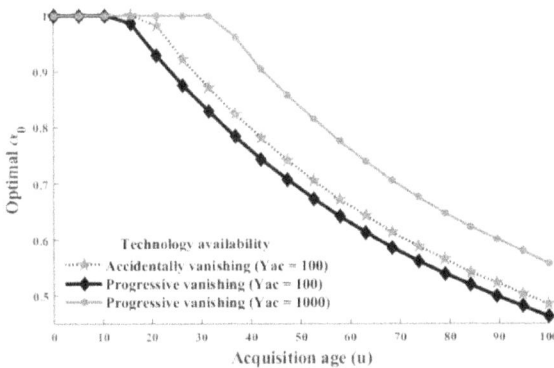

Figure 1.
Optimal rejuvenation ratio α_0.

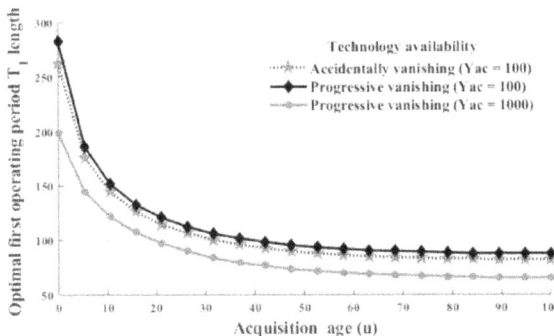

Figure 2.
Optimal operating period length T_1.

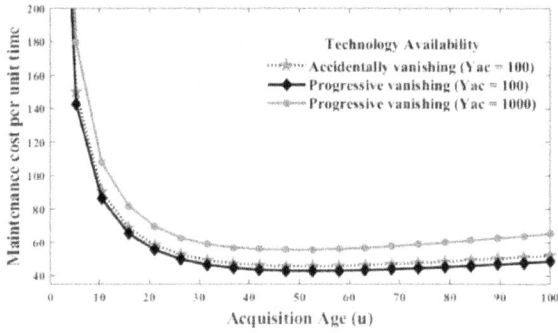

Figure 3.
Optimal maintenance cost per unit time.

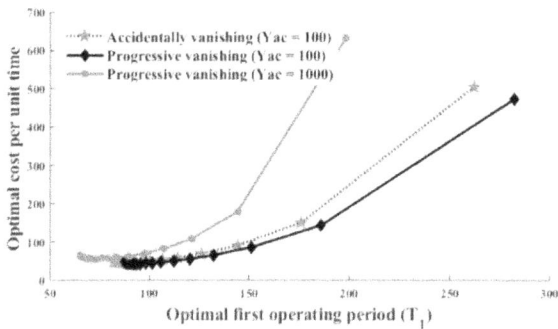

Figure 4.
Optimal period length T_k v.s maintenance cost per unit time.

of equipment with older technology remains expensive. Additional the uncertainty to predict the vanishing of technology in case of accidental makes the repairs performing uncertain and the cost per unit more expensive than progressive for a given age of technology. To finish we show through the **Figure 4** the evolution of optimal cost per unit time in function of operating period length. From **Figure 4**, we note that at optimal, the cost per unit time increases with the operating length. Therefore operating equipments with accidentally technology vanishing remains more expensive than progressive.

Table 2 resumes information and allows making comparison between parameters at optimal for each case. This table highlights some important information. Such as the use of equipments based on old technology remains expensive when the

Parameters	Accidental vanishing	Progressive vanishing	Progressive vanishing
	$Y_{AC} = 100$	$Y_{AC} = 100$	$Y_{AC} = 1000$
u	52.63	52.63	47.36
α_0	0.8054	0.6733	0.8577
T_1	87.67	93.27	73.66
$\tilde{C}(1)$	45.71	42.89	55.63
$C(1)$	4008.04	4001.04	4097.94

Table 2.
Optimal parameters for one operating cycle.

vanishing is progressive. In case of accidental vanishing process, the age of technology does not impact the optimal condition. However for the same age of technology, operating equipment with progressive vanishing is better than accidental because the prediction of pare parts availability remains easier for progressive than accidental vanishing technology.

5. Conclusion

This chapter analyzed an optimal maintenance policy strategy to operate second-hand equipment under uncertainty due to the indirect obsolescence concept. The uncertainty used, in the present work, is about the spare parts availability to perform some maintenance actions on equipment due to technology vanishing. The maintenance policy was defined and discussed with respect to several parameters relatively to equipment, technology and nature of repair. Therefore the optimality of maintenance policy was discussed and optimal conditions were derived mathematically. However finding these optimal parameters analytically remain complex. To reduce the complexity, a first optimal operating length was derived and compared according to the technology age and vanishing process. We highlight that the use of older technology requires less rejuvenation but remains more expensive than newer use. According to obtained results, we can note that the technology vanishing shows also the optimality of maintenance policy. Therefore the operating of equipment with technology whose vanishing process is accidental remains expensive. In the future works, we are going to focus on the development of algorithm to solve the complex optimization problem. This optimization will be made without restriction on the parameters.

Author details

Ibrahima dit Bouran Sidibe[1*†], Imene Djelloul[2†], Abdou Fane[1] and Amadou Ouane[3]

1 Centre de Formation et de Perfectionnement en Statistique (CFP-STAT), Bamako, Mali

2 Ecole Supérieure des Sciences Appliquées d'Alger (ESSA-Alger), Algiers, Algeria

3 Ministère de l'Enseignement Supérieur et de la Recherche Scientifique, Bamako, Mali

*Address all correspondence to: bouransidibe@gmail.com

† These authors contributed equally.

IntechOpen

References

[1] Barlow R, Hunter L. Optimum preventive maintenance policies. Operations Research. 1960;**8**(1):90-100 https://doi.org/10.1287/opre.8.1.90

[2] Barlow R, Proschan F: Mathematical theory of reliability. Society for industrial and applied mathematics. 1996. https://doi.org/10.1137/1.9781611971194

[3] Cho D, Parlar D: A survey of maintenance models for multi-unit systems. European Journal of Operation Research, 1991; 51: 1–23. https://doi.org/10.1016/0377-2217(91)90141-H

[4] Dekker R: Application of maintenance optimization models: review and analysis. Reliability Engineering and System Safety, 1996; 51 (3): 229–240. https://doi.org/10.1016/0951-8320(95)00076-3

[5] Jardin A, Tsang A: Maintenance, replacement and reliability, theory and applications. Taylor and Francis. 2006. https://doi.org/10.1201/b14937

[6] Nakagawa T, Mizutani S: A summary of maintenance policies for a finite interval. Reliability Engineering and System Safety, 2009; 94(1): 89–96. https://doi.org/10.1016/j.ress.2007.04.004

[7] Nakagawa T: Advanced reliability models and maintenance policies. Springer series in reliability engineering. 2006. https://www.springer.com/gp/book/9781848002937

[8] Lugtighei D, Jardine A, Jiangn X: Optimizing the performance of a repairable system under a maintenance and repair contract. Quality Reliability Engineering International, 2007; 23(1): 943–960. http://www.interscience.wiley.com/jpages/0748-8017/

[9] Chattopadhyay G.N, Murphy D.N.P: Warranty cost analysis for second-hand products. Mathematical and Computer Modelling, 2000; 31: 81–88. https://doi.org/10.1016/S0895-7177(00)00074-1

[10] Shafiee M, Finkelstein M, Chukova S: On optimal upgrade level for used products under given cost structures. Reliability Engineering & System Safety, 2011; 96: 286–291. http://www.sciencedirect.com/science/journal/09518320

[11] Khatab A, Diallo C, Sidibe I.B: Optimizing upgrade and imperfect preventive maintenance in failure-prone second-hand systems. Journal of Manufacturing Systems, 2017: 58–78. https://doi.org/10.1016/j.jmsy.2017.02.005

[12] Sidibe I.B, Khatab A, Kasambara A: Preventive maintenance optimization for a stochastically degrading system with a random initial age. Reliability Engineering & System Safety, 2017: 255–263. https://doi.org/10.1016/j.ress.2016.11.018

[13] Doyen L, Gaudouin O: Classes of imperfect repair models based on reduction of failure intensity or virtual age. Reliability Engineering & System Safety, 2004: 45–56. https://doi.org/10.1016/S0951-8320(03)00173-X

[14] Nawaz sharif M, Nazrul Islam M: The Weibull distribution as a general model for forecasting technological change. Technological forecasting and social change, 1980; 18: 247–256. https://doi.org/10.1016/0040-1625(80)90026-8

Reliability for Digital and Analog Systems

Chapter 5

Digital On-Chip Calibration of Analog Systems towards Enhanced Reliability

Michal Sovcik, Lukas Nagy, Viera Stopjakova
and Daniel Arbet

Abstract

This chapter deals with digital method of calibration for analog integrated circuits as a means of extending its lifetime and reliability, which consequently affects the reliability the analog electronic system as a whole. The proposed method can compensate for drift in circuit's electrical parameters, which occurs either in a long term due to aging and electrical stress or it is rather more acute, being caused by process, voltage and temperature variations. The chapter reveals the implementation of ultra-low voltage on-chip system of digitally calibrated variable-gain amplifier (VGA), fabricated in CMOS 130 nm technology. It operates reliably under supply voltage of $600\ mV$ with 10% variation, in temperature range from $-20°C$ to $85°C$. Simulations suggest that the system will preserve its parameters for at least 10 years of operation. Experimental verification over 10 packaged integrated circuit (IC) samples shows the input offset voltage of VGA is suppressed in range of $13\ \mu V$ to $167\ \mu V$. With calibration the VGA closely meets its nominally designed essential specifications as voltage gain or bandwidth. Digital calibration is comprehensively compared to its widely used alternative, Chopper stabilization through its implementation for the same VGA.

Keywords: on-chip digital calibration, PVT variations, aging compensation, reliability, input offset voltage, continuous operation, ultra-low voltage

1. Introduction

Every industry in our world shares the same fundamental motivation - packing as much functionality and power in the smallest possible size. This is naturally effective. In development of an integrated circuits fabrication technology, this trend is projecting in scaling down the minimum circuit element dimensions and circuit power supply voltage. In this way, higher performance and greater mobility is provided for electronic systems during their use. Such an advance, on the other hand, also introduces significantly increased random variations in circuit's electric specifications. The variations, in return, compromise the reliability already on the top level of electronic systems based on ICs as well as limit the functionality of circuits under constrained energy conditions. Deteriorated reliability of IC fabricated in modern nanotechnologies is significant not only between wafers and its series, but already within single die. Nowadays, ICs fabricated in $7\ nm$ or $5\ nm$

process nodes can still perform well in digital signal processing. However, their analog counterparts substantially suffer from impaired reliability yet in 130 nm technology. This is of high concern with precise circuits such as operational amplifier (OA), which is usually based on differential topology. Occurrence of any differences in its differential branches, which might be a result of random variations, simultaneously induce degradation of the amplifier key parameters. In terms of circuit electrical parameters, these variations are mirrored by transistor's threshold voltage (V_{TH}). The V_{TH} variance is commonly characterized by standard deviation in the matched transistor pair with respect to its size as follows:

$$\sigma(\Delta V_{TH}) = \frac{A_{VTH}}{\sqrt{W.L}},$$

(1)

where A_{VTH} is Pelgrom's technology coefficient, and W and L is the width and length of the transistor, respectively. The plot in **Figure 1** displays ΔV_{TH} for different process nodes according to works [1–5]. One can observe that downscaled process nodes suffer from much greater variance in V_{TH} with the change of device dimensions. In relative terms, the V_{TH} variance in 45 *nm* process node can reach 16% of mean value according to [6].

The adverse variations can be classified as rather acute - process, voltage and temperature variations, or they can occur after long term use as a consequence of electrical stress. The work [7] thoroughly analyzed the roots of these variations in a transistor as the fundamental IC element.

The following section describes the motivation behind our research. Section 3 explains the design details of the calibration methods implemented in this work. These include digital calibration (DGC) that was experimentally prototyped on a

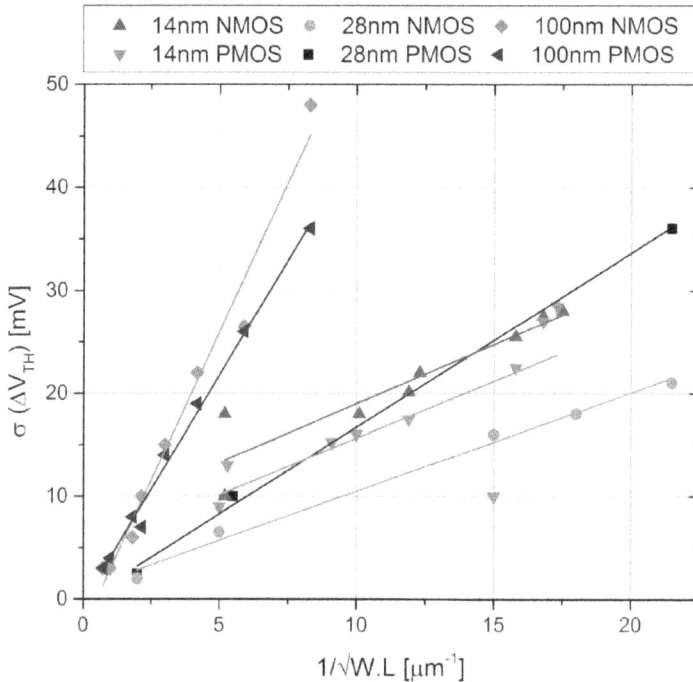

Figure 1.
Standard deviation of threshold voltage in matched transistor pair with respect to its size across different process nodes.

chip, its upgraded and optimized version (DGC2) and Chopper stabilization (CS) technique. Section 4 reveals results of extensive measurements and simulations of the proposed solutions. Finally, the Conclusion summarizes the most important outcomes of the presented research.

2. Motivation

Concluding the discussed state of the art, it is of high importance and it will be increasingly important in the future to compensate for stochastic variations in ICs to maintain reliable circuit operation. The parameter of an OA that is directly tied to any detrimental change in the transistor V_{TH} is the input offset voltage V_{INOFF}. It is therefore, an effective target of any calibration method. Among widely utilized techniques of IC calibration belong fuse trimming (one time or re-programmable), Chopper stabilization and auto-zero (AZ) technique.

In our work, we propose a promising alternative approach based on digital algorithm, that was utilized for the variable gain amplifier. It is fully integrated on-chip system, which is potentially entirely autonomous. This method creates moderate area and energy consumption overhead, but it preserves frequency performance of the calibrated VGA. The methods of calibration substantially differ in fundamentals of operation, which makes the comparison between them difficult. Therefore, we implemented also the CS method for the same VGA. In this way, we can precisely compare the digital calibration to an alternative solution in terms of implementation details.

3. Implementation of calibration methods

This section proposes an insights into design fundamentals of implemented methods of calibration. The fabricated system of DGC, its optimized version for continuous systems and CS will be analyzed. All methods are designed consistently in order to achieve a clear and relevant comparison. They are implemented in standard CMOS 130 nm process node and operate under the supply voltage of 600 mV. Each method was utilized for the same VGA, which was previously realized in the same technology. The calibration is based on compensation of V_{INOFF} the designed and manufactured VGA exhibits. The VGA is nominally designed to reach DC gain magnitude of 33 dB with bandwidth of 17 kHz with capacitance load of 10 pF.

3.1 Fabricated system of DGC

Figure 2 depicts the block structure of digitally calibrated VGA, which has already been fabricated and evaluated. The calibration subcircuit connects to VGA through sensing ($P_{S1,2}$) and compensation ($P_{C1,2}$) ports. These are followed by the control and compensation blocks of the calibration subcircuit.

The control block senses the actual V_{INOFF} through the voltage comparators, as it is tied to output offset voltage with the following formula:

$$V_{OUTOFF} = A_{CLG}.V_{INOFF}, \tag{2}$$

where V_{OUTOFF} is the output offset voltage and A_{CLG} is amplifier's closed-loop gain.

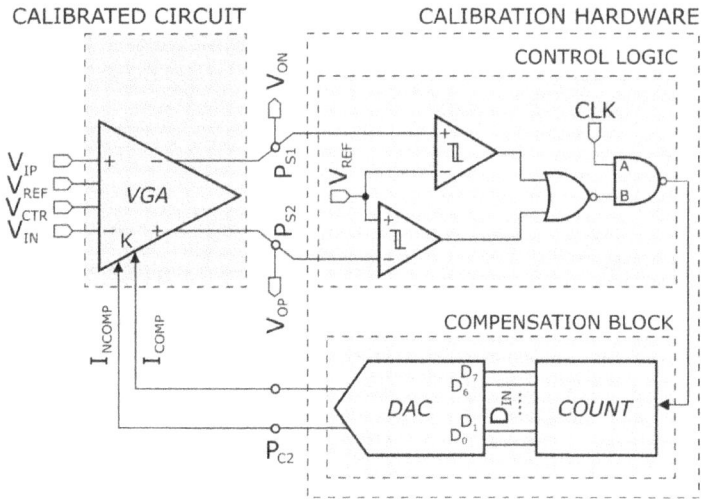

Figure 2.
Implementation of digital calibration for VGA input offset voltage cancelation.

The compensation block is connected to VGA topology so it can modify the current flow in its differential branches. As long as VGA outputs ($V_{OP,ON}$) do not cross the reference voltage (V_{REF}) at comparators' inputs, the digital-to-analog converter (DAC) adjusts the currents in the VGA. These currents are proportionally projected to V_{INOFF}. When either one of VGA outputs crosses the V_{REF}, the corresponding comparator terminates the clock signal, which drives the compensation block. In this way, V_{OP} and V_{ON} are brought close to each other, diminishing the offset voltage.

The direct schematic detail of compensation port is depicted in **Figure 3**. In this configuration, the DAC output current is mirrored to VGA through the active load of the amplifier, transistors M_{P1} and M_{P2}. The current mirror is formed by transistors M_{Z2}, M_{Z3}, M_{Z5} and M_{Z6}. The current mirror on the side of VGA is connected in bulk-driven configuration, which is specific by using the substrate electrodes to control the transistor. The reason for using this topology is better matching in drain-source voltage of mirroring devices M_{Z5} and M_{P1} (analogically M_{Z6} and M_{P2} for the other VGA branch), as their gate and substrate electrodes are correspondingly tied together. Also the bulk-driven topology performs better in low-voltage conditions

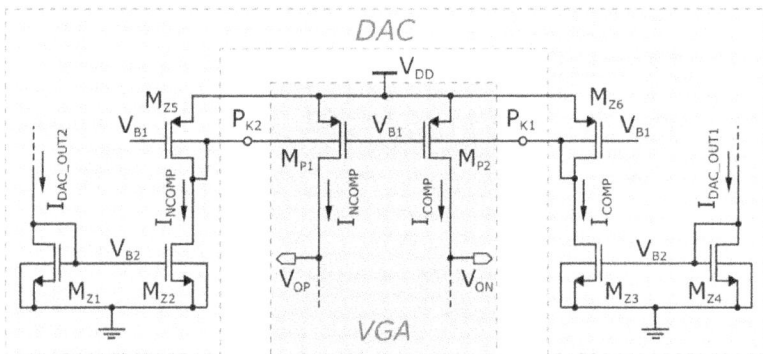

Figure 3.
The detail of compensation port at schematic level. This represents interface between DGC and VGA.

of the designed system, as it allows to modify the. This is clear from the following formula:

$$V_{TH} = V_{TH0} + \gamma \left(\sqrt{|2\Phi_F + V_{SB}|} - \sqrt{|2\Phi_F|} \right), \tag{3}$$

where V_{SB} is the bulk electrode voltage, V_{TH0} is threshold voltage at $V_{SB} = 0V$, γ is the body effect coefficient and Φ_F is the Fermi potential.

The layout of complete DGC system is depicted in **Figure 4**. The main blocks are highlighted from the principle structure point of view from **Figure 2**. The dimensions are shown in μm. The VGA covers 18500 μm^2, while the DGC circuits together take 19000 μm^2 without the optional trimming fuses.

Figure 5 displays the micrograph of an experimental chip prototype, where the digital calibration system is marked.

3.2 Optimized system of DGC for continuous operation

As it was described in the previous section, the voltage–current conditions in VGA are modified through the iterations of calibration cycle. The VGA is imbalanced in this process and therefore, cannot perform its intended function.

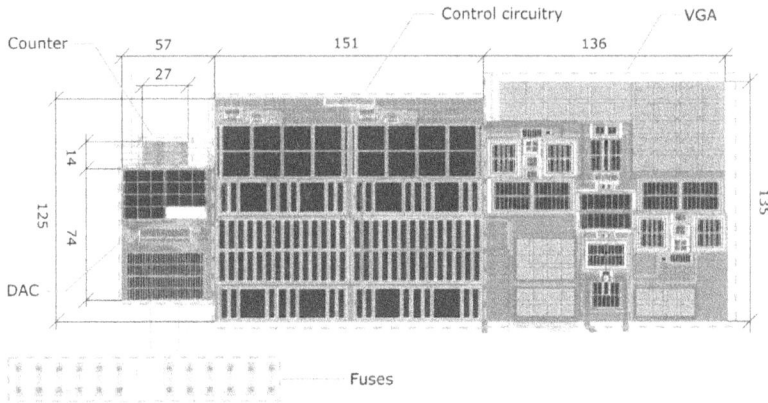

Figure 4.
The layout of the whole calibration system (dimensions in μm).

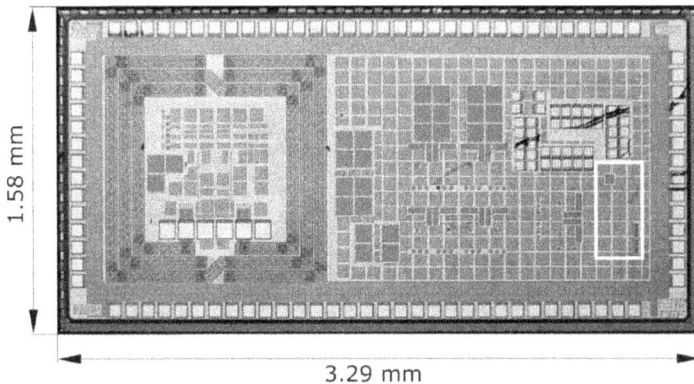

Figure 5.
The micrograph of die with marked area of implemented DGC.

In systems without the need to continuously process the signal, this does not represent an issue, since through proper synchronization the calibration cycle it can be carried out during idle phase. To utilize the digitally calibrated VGA in an application requiring continuous operation, the technique called "Ping-pong" facilitates the necessary continuity [8, 9]. Its principle can be illustrated by the block diagram shown in **Figure 6**. It uses the calibrated VGA in two identical versions, where the one (VGAX) provides the amplified output while the other (VGAY) is calibrated. When the circuit characteristics change due to temperature or voltage variations, both amplifiers switch their roles. In this way, the output is provided only by already calibrated VGA.

To explain the operation of "Ping-pong" digital calibration (PDGC) more clearly, **Figure 7** shows exemplary transient flow of the operation on key signals of system in **Figure 6**. The temperature change in this example, creates the conditions, in which the calibration is required. During the system initialization, the VGAX is calibrated and its output offset voltage ($V_{OUTOFFX}$) is suppressed to a range of hysteresis ΔV_{HYST}. The overall system output (V_{OUT}) is consequently switched to VGAX in time t_0. When the V_{INOFFX} drifts above ΔV_{HYST} with temperature, the calibration of VGAY begins in time t_1. When the calibration cycle is completed, the signal SW_Y indicates it by a rising edge and the output of VGAY is switched to V_{OUT} in time t_2. In this way, the phases of PDGC are autonomously interchanged according to temperature, which is evaluated by a temperature sensor. The maximum overall output offset voltage at V_{OUT} is maintained nearly above ΔV_{HYST} without any substantial corruptions.

During switching VGA versions, the V_{OUT} can suffer from voltage spikes. These produce significantly smaller artifacts of frequency spectrum than spikes resulting from charge injection present in switched calibration methods such as auto-zero technique. Also the spikes can be eliminated by incorporating transfer phase, when the calibrated amplifier is firstly connected as a buffer before being fully switched to the system output [8].

Utilization of "Ping-pong" technique required a few modifications to a system of digitally calibrated VGA in **Figure 2**. These are incorporated in its upgraded and optimized version, depicted in **Figure 8** with more insights into the control logic block shown in **Figure 9**. In this modified system, the power-on-reset (POR) circuit is present, which is controlled by *RPT* pin. The POR activates the calibration of idle

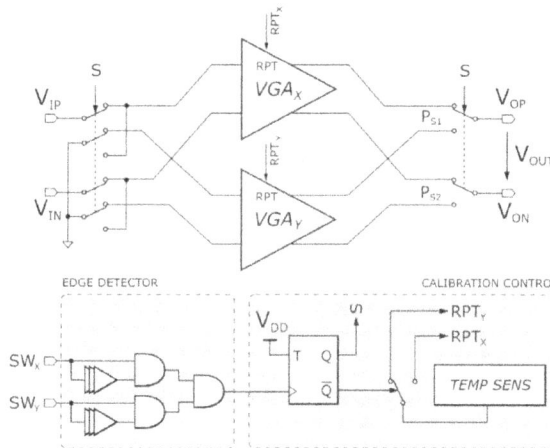

Figure 6.
The "ping-pong" technique for digitally calibrated VGA with continuous output.

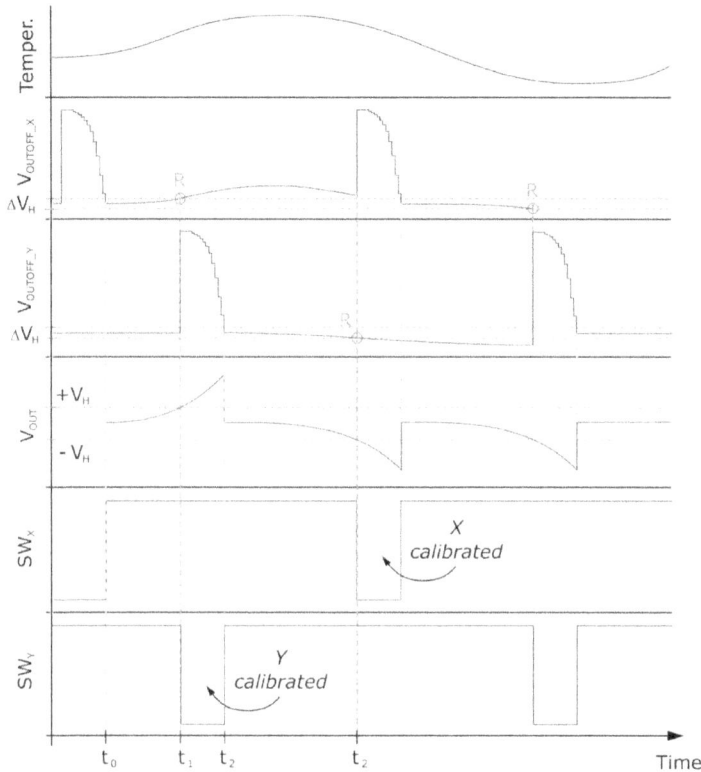

Figure 7.
The "ping-pong" technique for digitally calibrated OA with continuous output.

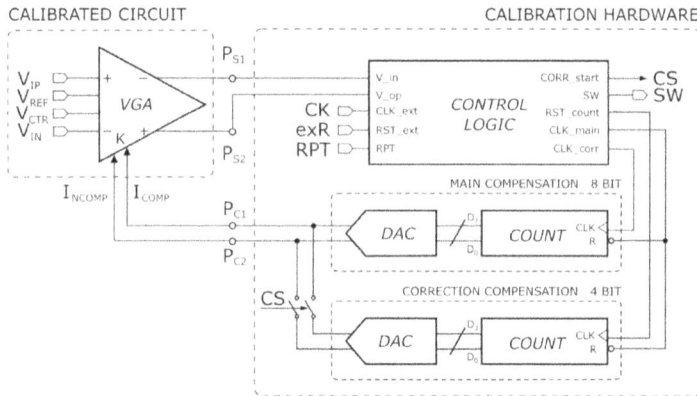

Figure 8.
The improved version of digitally calibrated VGA for integration into "ping-pong" method.

VGA, when it is required to be switched to the system output. Also the pin SW indicates the calibration cycle is completed. The new version of DGC depicted in **Figure 8** also incorporates modifications, which brings better overall efficiency of the discussed method. There is an additional 4-bit DAC, which corrects the residual offset remaining after the main calibration cycle has been completed. In this way, the magnitude of offset voltage is further suppressed. The two sensing comparators of the previous version are now replaced by a single one, which is much simpler

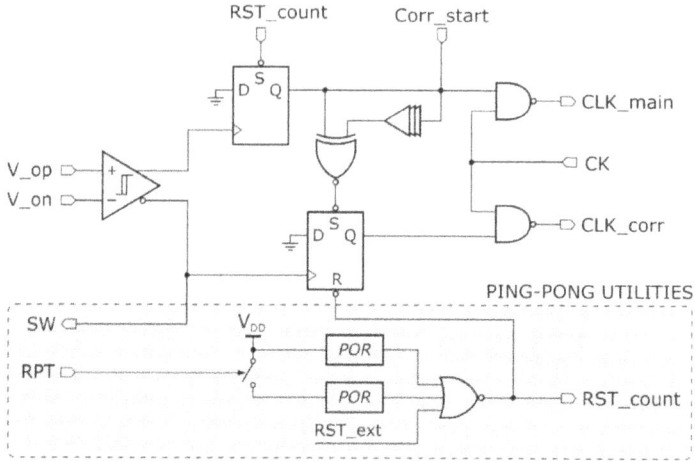

Figure 9.
Optimized control block for digital calibration with upgrades for "ping-pong" approach.

solution but with tighter specifications for a successful calibration. A D-type flip-flops in the control block terminate the distribution of *CLK* signal when the calibration is finished, which prevents unintended launch of another calibration cycle.

The optimized DGC from **Figure 8** covers only 6000 μm^2 instead of 19000 μm^2 taken by the original version. Also the power consumption of calibration circuits is lowered from $41\mu W$ to just $6\mu W$, which is almost 7 times lower.

3.3 Chopper stabilization

In order to obtain precise comparison of calibration methods, so-called chopper stabilization has been utilized for the same VGA as in DGC, serving as an alternative method. In this way, the figure of merit (FOM) can be defined by considering the chip area coverage, power consumption and residual V_{INOFF}, as main parameters obtained over sufficient number of samples. The proposed chopper stabilization has been build according to the basic principle, without complex techniques, which would require an auxiliary amplifiers.

The block diagram of chopper-stabilized VGA is shown in **Figure 10**. It consists of modulator and demodulator employed at the VGA input and output, respectively. These are switched by signal $m(t)$ with frequency of 20 *kHz* so that it falls in VGA bandwidth. In order to make the comparison to DGC objective and relevant, we established the limit for total harmonic distortion (THD) at the value of 1%. The output of demodulator provides the carrier signal $m(t)$ with the envelope formed by V_{IN}. Therefore, the low-pass filter needs to be employed at each output channel. As there are requirements for low THD and low corner frequency (f_c) of 10 *kHz* (suppressing $m(t)$), the filter must incorporate significantly large values of capacitor

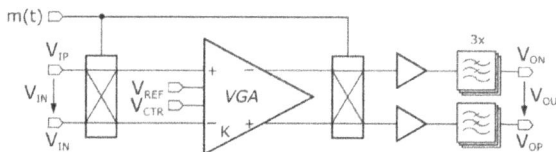

Figure 10.
The block diagram of chopper-stabilized VGA.

and resistor. Moreover, the f_c needs to be as close as possible to $m(t)$ frequency, so that the filter cuts the smaller portion of overall VGA bandwidth. To meet this requirement, the filter needs to reach significant steepness or in other words, the order. The mentioned specifications demand rather complex solution consisting of a frequency filter with large values of passive components. It was realized as Sallen-Key active filter of 2nd order, which is used 3 times in series for each VGA output channel. These two pairs of the filter cover together approximately 800 000 μm^2. While the VGA occupies only 18 500 μm^2, using the CS in this conditions would create a tremendous area overhead.

The large silicon area covered by frequency filters can be mitigated by switched capacitors that are used instead of classic resistors. On the other hand, the switching frequency in this case, would demand yet another design solution to maintain a low distortion of the processed input signal.

4. Verification

Fabricated system of DGC, described in section 3.1 was experimentally verified. The measurements have been carried out on 10 packaged IC samples at the ambient temperature of 27°C. All implemented methods of calibration for VGA were extensively verified through simulations in Cadence environment. These included process corner analysis, Monte Carlo (MC) and RelXpert reliability analysis. Each simulation type has been carried out in the temperature range from −20°C to 85°C. MC was performed with 150 sample scenarios. Reliability analysis simulates the operation of an IC after initial electrical stress and operation after 10 years of stress. It is based on the bias temperature instability (BTI) and hot carrier injection (HCI), the phenomena which gradually degrade the IC reliability and dependability. It does not consider time dependent dielectric breakdown since this is rather acute in nature. Reliability analysis considered simultaneously the process corners and geometry mismatch as well.

In order to optimally design the calibration circuits, the experimental measurements of V_{INOFF} in the same VGA without calibration (previously fabricated) hardware were performed. Measurements were carried out on 60 naked dies. The resulting offset distribution is displayed in **Figure 11** and compared to MC result distribution. 3σ range of V_{INOFF} reaches approximately 10 mV. Hypothetically, with

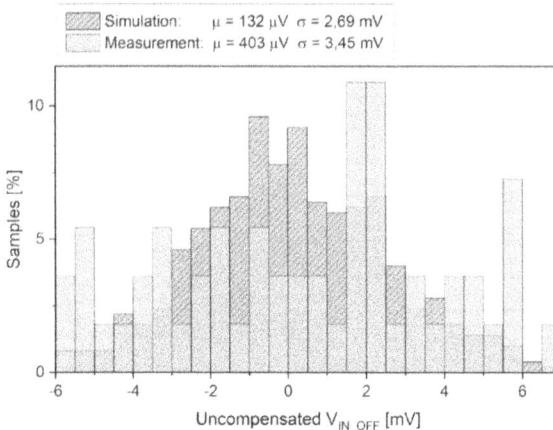

Figure 11.
Distribution of V_{INOFF} of uncalibrated VGA. Comparison of experimental measurements and MC analysis.

the VGA nominal gain of 33 dB such V_{INOFF} would project to 460 mV in the differential output voltage. Considering the VGA being supplied by 600 mV, its outputs would reach almost to the supply rails range even in the optimum operating point.

Histograms depicted in **Figures 12–14** compare the MC simulation results of V_{INOFF} of digitally calibrated VGA, chopper-stabilized VGA and VGA without a calibration. The achieved results prove a comparable performance of DGC and CS in terms of standard deviation, while CS reaches better centering of distributions towards 0 V. Both methods of calibration successfully suppress the V_{INOFF} in magnitude of orders through the whole temperature range.

Figure 15 displays the transient flow of digital calibration cycle through the VGA output voltages. It compares the best and worst case scenarios of technology process corners after the initial stress and after 10 years of operation. After the calibration is successfully completed, the VGA is fed the harmonic signal of 0.5 mV differential amplitude and frequency of 1 kHz. The detail of plot in right-hand side part of **Figure 15** zooms the differential output voltage of the calibrated VGA in proper operation. The calibration cycle was controlled by clock signal with frequency of 1 kHz. **Figure 16** displays the measured flow of calibration cycle also through VGA

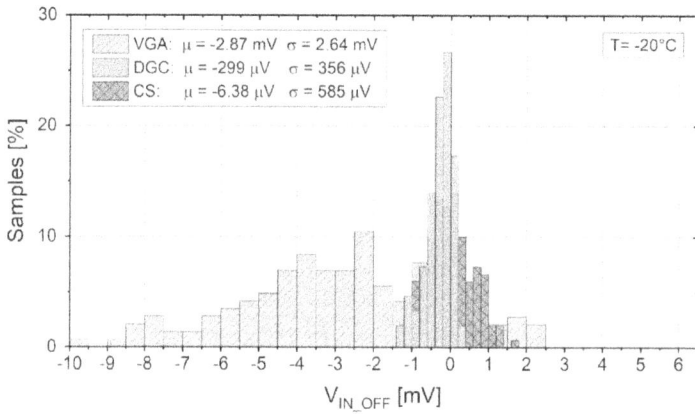

Figure 12.
MC simulation results of V_{INOFF} in digitally calibrated VGA, chopper stabilized VGA and VGA without calibration. Temperature of −20 C.

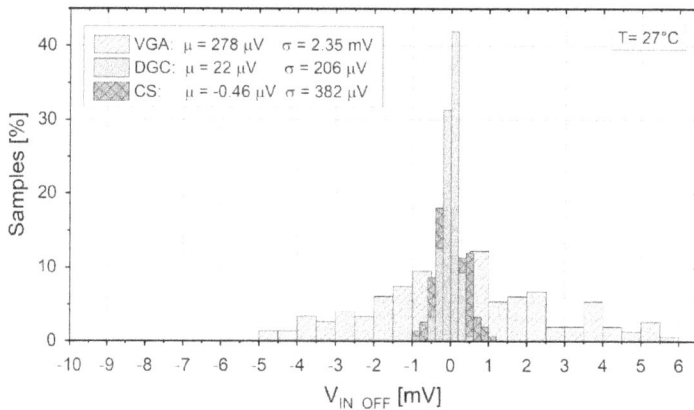

Figure 13.
MC simulation results of V_{INOFF} in digitally calibrated VGA, chopper stabilized VGA and VGA without calibration. Temperature of 27°C.

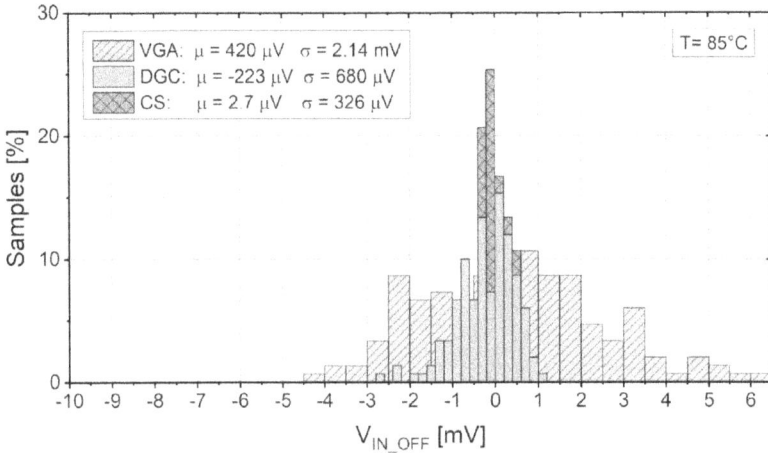

Figure 14.
MC simulation results of V_{INOFF} in digitally calibrated VGA, Chopper stabilized VGA and VGA without calibration. Temperature of 85°C.

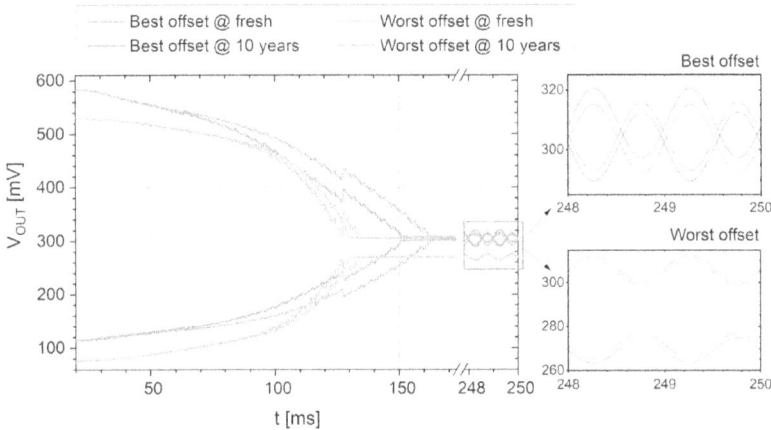

Figure 15.
Reliability analysis of digitally calibrated VGA considering process corners and geometry mismatch. The calibration cycle at fresh IC is compared to the IC of 10 years of age.

output voltages. The best and worst cases of the measurement and simulation results are compared. The calibration was performed with CLK frequency of 350 *kHz*. The whole cycle from beginning to the point where VGA is ready for proper operation lasts between 210 *μs* and 319 *μs*. **Figure 17** compares the gain of VGA obtained by measurement and simulation in the best and worst cases. Measurements were carried out using the maximum input signal amplitude for VGA, which allowed to keep the total harmonic distortion under 1%. As one can observe, the experimental results fit the simulations very well. The upgraded, more precise DGC version, described in section 3.2 is actually being re-designed at the moment. **Figures 18–20** compares the MC simulation results of V_{INOFF} to the prototype version over the industrial temperature range. The mean value of offset is compensated in one order of magnitude further towards 0 *V* by means of new DGC. Standard deviation remains approximately constant. The reason behind it is supposed to be the offset voltage of sensing comparator in DGC. It causes premature termination of main calibration cycle, which alone produces excessive residual V_{INOFF}. Consequently, the issue is amplified

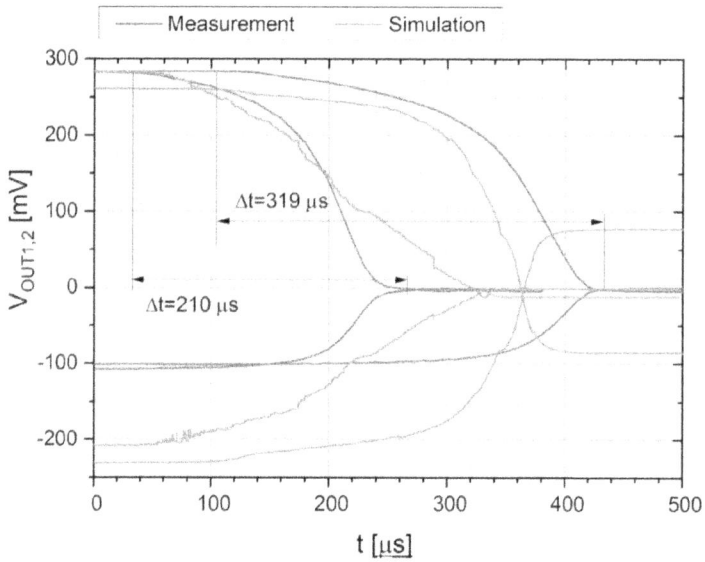

Figure 16.
Transient flow of the calibration cycle. The comparison of the best and the worst case of measurement and simulation.

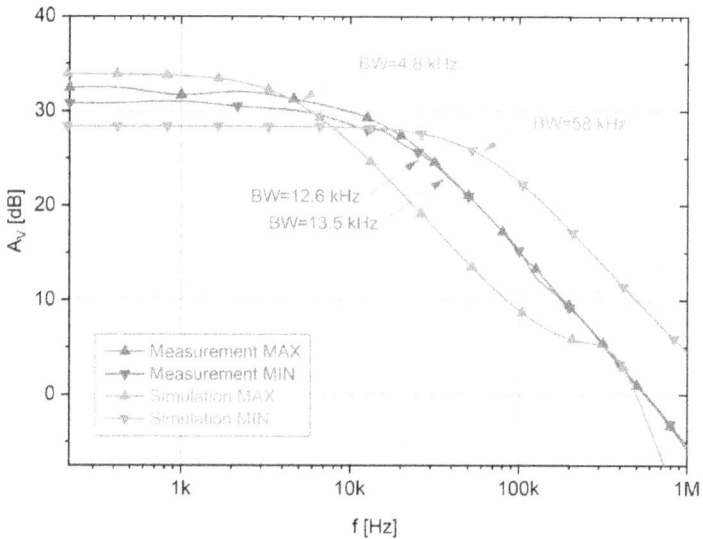

Figure 17.
Measured vs. simulated gain of the digitally calibrated VGA.

because faulty termination launches correction calibration cycle in the wrong direction, enhancing the residual V_{INOFF} even further. On the other hand, this shortcoming of upgraded DGC can be easily resolved by auto-zeroing the sensing comparator. As it processes discontinuous signal (proportional to DAC output), the proper synchronization of switching the AZ sample phase with CLK of calibration would preserve the overall DGC progress untouched.

Table 1 summarizes the most important results of calibration methods, implemented in this work. There are compared measurement and simulation results

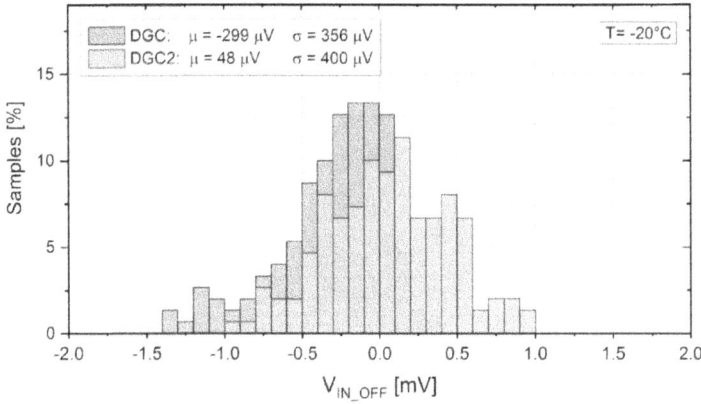

Figure 18.
Measured vs. simulated gain of the digitally calibrated VGA for temperature of −20°C.

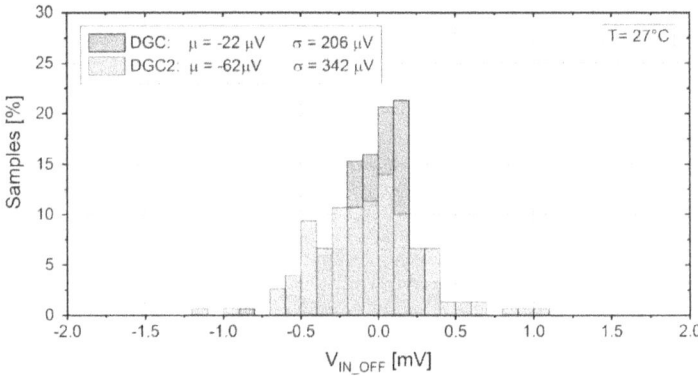

Figure 19.
Measured vs. simulated gain of the digitally calibrated VGA for temperature of 27°C.

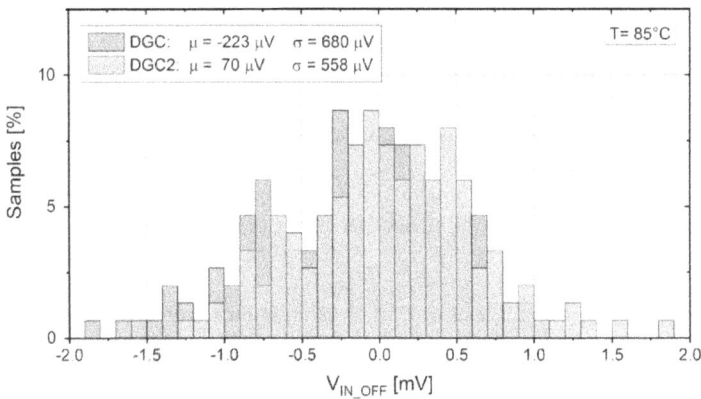

Figure 20.
Measured vs. simulated gain of the digitally calibrated VGA for temperature of 85°C.

of prototype DGC version and simulation results of upgraded DGC (DGC2) and CS. The whole data were obtained at room temperature of 27°C for the purpose of comparison to related works shown in **Table 2**. As all details in these designs were

		DGC				DGC2[3]		CS	
		min	max	μ	σ	μ	σ	μ	σ
Year					2020				
Node	[nm]				130				
V_{DD}	[V]				0.6				
S_{OA}[1]	$[.10^3 \mu m^2]$				18.4				
S_{CH}[2]	$[.10^3 \mu m^2]$		19			6		800	
P_{OA}[1]	[μW]	—	—	38	3	38.5	3.6	30.4	3.1
P_{CH}[2]	[μW]	—	—	41	6.1	5.6	0.7	62.7	11
V_{INOFF}	[μV]	13	167	22	206	18.5	390	2.75	95
THD	[%]		1	0.04	0.02	0.83	0.07	0.05	0.03
A_{DC}	[dB]	31	33	35.3	2.7	33.6	8.2	30.23	2.2
BW	[kHz]	12.6	13.5	15.25	2.2	15.9	2	0.9	—
GBW	[MHz]		0.6	0.75	0.06	800	61	3	—
FOM	[−]		—	**41**		**340**		**4**	
Results		Meas.				Sim.			
Calibration					Dig.			Chop.	

[1] Chip area coverage (S) and power consumption (P) of the calibrated OA.
[2] Chip area coverage (S) and power consumption (P) of the calibration hardware.
[3] Upgraded and optimized version of the prototyped DGC.
Bold values are emphasize the parameter FOM, which is one of most important results in this work.

Table 1.
The main results of the proposed calibrated systems.

accessible, the figure of merit (FOM) has been determined according to the following formula:

$$FOM = \frac{1000}{\mu(V_{INOFF}).A.P} \tag{4}$$

where the parameters are:

- $\mu(V_{INOFF})$ - mean value of residual input offset voltage after calibration over sufficient number of samples,

- $A = A_{CH}/A_{CC}$ - ratio of calibration hardware die area versus area of calibrated circuit,

- $P = P_{CH}/P_{CC}$ - power consumption of calibration hardware versus power consumption of the calibrated circuit.

The FOM value evaluates the power and area efficiency of the obtained calibration result over a sufficient number of samples. Its coefficients are in denominator, and therefore, the greater value of FOM proves better overall performance of calibration method. The value of coefficients is multiplied by 1000 in numerator to shift the results into the order of tens to thousands. The crucial condition for objectivity in this comparison is the THD of the amplifier is maintained under 1% at any method of calibration.

		This work		[11]	[12]	[13]	[14]	[10]
		μ	σ					
Year		2020		2015	2008	2008	2010	2013
Node	$[nm]$	130		130	350	180	130	130
V_{DD}	$[V]$	0.6		1.2	1	1.8	1.2	2.8
S_{OA} [1]	$[.10^3 \mu m^2]$	18.4		12.3	224	36	34	630
S_{CH} [2]	$[.10^3 \mu m^2]$	6		12.3	17.4	30	33.2	261
P_{OA} [1]	$[\mu W]$	38.5	3.6	13920	7900	11000	**2000**	156800
P_{CH} [2]	$[\mu W]$	5.6	0.7				2400	
V_{INOFF}	$[\mu V]$	62	342	5	σ=538	126	75	0.097
THD	[%]	0.83	0.07	—	0.01[3]	0.4	—	—
A_{DC}	$[dB]$	33,6	8,2	0–60	−22 - 30	−6 - 58	32	90
BW	$[kHz]$	15.9	2	250000	17000	22000	57000	40000
OA stages		1		4	3	3	1	1
FOM	[−]	**340**		205[†]	24	10	11[†]	25 000[†]
Results source		Sim.		Meas.	Sim.	Meas.	Meas.	Meas.
Calibration method		Dig.		Analog.	Analog.	Dig.	Analog.	Dig.

[1] *Chip area coverage (S) and power consumption (P) of the calibrated OA.*
[2] *Chip area coverage (S) and power consumption (P) of calibration hardware.*
[†] *These works does not contain information about THD, therefore the comparison is not fully relevant.*
Bold values are emphasize the parameter FOM, which is one of most important results in this work.

Table 2.
The results comparison of the proposed work to related works.

It is obvious that CS outperforms even DGC2 in term of V_{INOFF}. It maintains the nominally designed low-frequency gain (A_{DC}) at 30.23 dB. On the other hand, the bandwidth (BW) is seriously constrained due to extensive filtering and the principle of CS alone.

However, this is not an issue with DGC and DGC2 approach, as the simulation results of BW after the calibration converge on the nominally designed 17 kHz. The measurement value of BW exhibits a certain decline, which could be assigned to

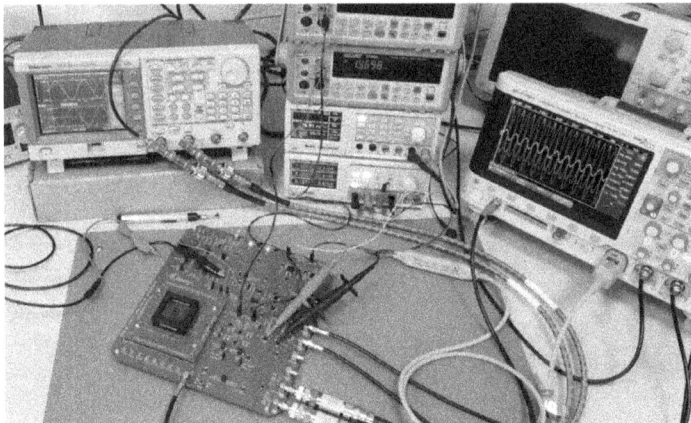

Figure 21.
The automated measurement setup for evaluation of the prototype calibrated systems.

parasitic capacitances associated with measurement setup, as well as to the measurement error due to small signal amplitude. Thanks to optimization of occupied silicon area and power consumption the DGC2 outperforms CS according to FOM. **Table 2** compares the proposed results to related works in the scientific area. This comparison was problematic as the calibration methods are most often built in more complex systems as DBS receiver circuit [10] and others. They are therefore side topics and authors do not mention the details, which are critical in terms of calibration. One such limitation is missing the level of THD, which sets equal conditions with calibration methods, presented in this chapter. The FOM values proves DGC2 for being a competitive solution for calibration of analog integrated circuits.

Figure 21 depicts the setup for automated measurements, which served for the evaluation of prototype chips.

5. Conclusions

Prevalent trends in IC fabrication reveal an increasing sensitivity to process, voltage, temperature variations and aging that projects directly into degradation of the overall perfomance of integrated systems. While digital systems are proven to be robust enough to above mentioned fluctuations even below 10 *nm* process node, their analog counterparts suffer from significantly decreased yield already in 130 *nm* technology node. This is confirmed by an actual research and also by experimental results provided within this chapter. Various methods of calibration have been presented and proven to effectively aid analog integrated circuits in preserving their reliability. These methods are rather complex, which constricts them for very specific use. This chapter extensively analyses the digitally controlled calibration method aimed at compensating the input offset voltage of the variable-gain amplifier. Chopper stabilization technique was implemented for the same amplifier, ensuring an objective comparison. Digital calibration was verified by simulations and experimental measurements on a prototype chip, which uniformly proved the reliable performance of the calibrated system. The established figure of merit shows that digital calibration represents an effective solution for preservation of reliability level in continuous systems with low distortion. The proposed method operates with lowest supply voltage level and also achieves the lowest silicon area and power consumption between the compared solutions, while maintaining competitive level of the calibrated input offset voltage. The area, which is added by calibration hardware can be mitigated by relieving demands for robustness of the calibrated circuit. By means of digital processing the method itself is robust against electrical variations.

Concluding the provided results, the digital calibration proves to be a promising solution in aiding the analog ultra-low voltage systems on chip towards a reliable operation, which is enormous challenge in modern nanoscale technologies. The proposed method is area and power efficient, while its operation remains stable over at least 10 years life span. It can be easily integrated along the calibrated circuit on a single chip in deep sub-micron process nodes. Taking into account the further research, it can become fully autonomous. In this way, the handling of calibrated circuit remains intact of calibration management.

Acknowledgements

This research was supported in part by the Slovak Research and Development Agency under grant APVV-19-0392, and ECSEL JU under project PROGRESSUS (Agr. No. 876868).

Author details

Michal Sovcik, Lukas Nagy, Viera Stopjakova* and Daniel Arbet
Slovak University of Technology, Bratislava, Slovakia

*Address all correspondence to: viera.stopjakova@stuba.sk

IntechOpen

References

[1] Z. Al Tarawneh, *The effects of process variations on performance and robustness of bulk CMOS and SOI implementations of C-elements*. PhD thesis, Newcastle University, 2011.

[2] B. C. P. et al, ""variability evaluation of 28nm fd-soi technology at cryogenic temperatures down to 100mk for quantum computing," in *Technical Highlights from the 2020 Symposia on VLSI Technology and Circuits [to be published]*, 2020.

[3] G. Angelov, D. Nikolov, M. Spasova, and R. Rusev, "Study of process variability-sensitive local device parameters for 14-nm bulk finfets," in *2020 43rd International Spring Seminar on Electronics Technology (ISSE)*, pp. 1–4, 2020.

[4] F. A. et al., "Low leakage and low variability ultra-thin body and buried oxide (ut2b) soi technology for 20nm low power cmos and beyond," in *2010 Symposium on VLSI Technology*, pp. 57–58, 2010.

[5] H. Edwards, T. Chatterjee, M. Kassem, G. Gomez, F.-C. Hou, and X. Wu, "Device physics origin and solutions to threshold voltage fluctuations in sub 130 nm cmos incorporating halo implant," 10 2010.

[6] M. Onabajo and J. Silva-Martinez, *Analog circuit design for process variation-resilient systems-on-a-chip*. Springer Science & Business Media, 2012.

[7] D. Wolpert and P. Ampadu, *Managing Temperature Effects in Nanoscale Adaptive Systems*. Springer New York, 2011.

[8] M. Pastre and M. Kayal, *Methodology for the digital calibration of analog circuits and systems*. Springer, 2006.

[9] M. Kayal, R. T. L. Saez, and M. Declercq, "An automatic offset compensation technique applicable to existing operational amplifier core cell," in *Proceedings of the IEEE 1998 Custom Integrated Circuits Conference (Cat. No.98CH36143)*, pp. 419–422, May 1998.

[10] S. Li, J. Li, X. Gu, H. Wang, M. Tang, and Z. Zhuang, "A continuously and widely tunable 5 db-nf 89.5 db-gain 85.5 db-dr cmos tv receiver with digitally-assisted calibration for multi-standard dbs applications," *IEEE Journal of Solid-State Circuits*, vol. 48, pp. 2762–2774, Nov 2013.

[11] Y. Zhang, Y. Fei, Z. Peng, and F. Huang, "A 250mhz 60db gain control range 1db gain step programmable gain amplifier with dc-offset calibration," in *2015 International Symposium on Intelligent Signal Processing and Communication Systems (ISPACS)*, pp. 227–230, Nov 2015.

[12] P. Mak, S. U, and R. P. Martins, "On the design of a programmable-gain amplifier with built-in compact dc-offset cancellers for very low-voltage wlan systems," *IEEE Transactions on Circuits and Systems I: Regular Papers*, vol. 55, pp. 496–509, March 2008.

[13] Z. Cheng and J. Bor, "A cmos variable gain amplifier with dc offset calibration loop for wireless communications," in *2006 International Symposium on VLSI Design, Automation and Test*, pp. 1–4, 2006.

[14] Xiaojie Chu, Min Lin, Zheng Gong, Yin Shi, and Fa Foster Dai, "A cmos programmable gain amplifier with a novel dc-offset cancellation technique," in *IEEE Custom Integrated Circuits Conference 2010*, pp. 1–4, 2010.

www.ingramcontent.com/pod-product-compliance
Lightning Source LLC
Chambersburg PA
CBHW081236190326
41458CB00016B/5809